Frank J. Firth, O. H. Jewell Filter Company

Purification of City Water Supplies by Sand Filtration

Frank J. Firth, O. H. Jewell Filter Company

Purification of City Water Supplies by Sand Filtration

ISBN/EAN: 9783743423558

Manufactured in Europe, USA, Canada, Australia, Japa

Cover: Foto ©berggeist007 / pixelio.de

Manufactured and distributed by brebook publishing software
(www.brebook.com)

Frank J. Firth, O. H. Jewell Filter Company

Purification of City Water Supplies by Sand Filtration

O. H. Jewell Filter Company,

(INCORPORATED.)

O. H. JEWELL, C. & M. E., ... President.
W. B. BULL, 1st Vice President.
IRA H. JEWELL, . . . 2d Vice President.
WM. M. JEWELL, Ph. G., Sec'y and Treas.
W. WAGNER, C. E., . Superintendent.
GEO. F. HODKINSON, Sales Agent.

OFFICE AND FACTORY:

73-75 West Jackson Street, Chicago.

Cable and Telegraphic Address: "JULFILTER," CHICAGO.

The Morison=Jewell Filtration Co.

(INCORPORATED.)

HENRY LEVIS, ... President.
S. L. MORISON, Vice Pres. and Gen'l Mgr.
W. S. WYLIE, . . Treasurer.
AMORY A. DIXON, . . . Secretary.
C. R. DAVIS, Superintendent.

Cable Address,
"LAYSTALL," New York.

PHILADELPHIA, 26 South Fifteenth St.
NEW YORK, . 1310 Havemeyer Building.
NEW JERSEY, 202 Market St., Camden.

W. P. Dunn Company, Printers,
187 Adams Street,
Chicago.

GENERAL NOTES.

MANUFACTURING.

OUR factory is the largest if not the only one in the world exclusively devoted to the manufacture of filters. It is equipped with all the latest improved machinery and with many special machines designed especially for our work. None but skilled workmen are employed, several ot whom have been in our employ for many years.

DRAFTING.

The drafting and designing department is most complete. Drawings and preliminary sketches showing the general arrangement will be made to accompany propositions when desired, provided we receive the necessary data concerning the proposed location, etc. No charge will be made for these drawings.

Complete working plans of large gravity and pressure filter plants are made after acceptance of proposition and submitted for approval. For this purpose our engineer generally visits the works to obtain measurements and full information in detail.

LABORATORY.

We have a skilled chemist in charge of our laboratory, where all of our own chemical, microscopic, and bacteriological tests of waters are made. It has been our custom for years to make complete analyses and practical tests of all waters, unless we are perfectly familiar with them, to determine the best and most economical methods of purification. Parties wishing complete reports on the purification of their water supplies should write for instruction pamphlet before sending sample.

PATENT PROTECTION.

The Jewell filters and auxiliary appliances are the result of over twenty years' experience and practical work in the purification of waters, during which time nearly 100 patents have been issued thereon, in this and foreign countries, and our customers are amply protected in their use of any apparatus or appliance purchased of us.

PRICES.

Owing to the various conditions attending the location and erection of our filters, especially for city water works, paper mills, sugar refineries, and other large industries, where generally two or more of them are connected in a battery, the arrangement of each plant must be separately designed. We are, therefore, prepared to make an estimate of cost only when we have become fully conversant with the circumstances bearing directly on each individual case; and we have, therefore, omitted all figures of cost in this issue. **Our prices are most reasonable, and unquestionably the lowest for the highest possible standard of workmanship, efficiency and durability.**

Facts in Filtration.

The following facts should be known and always remembered:

(1) All filtration is sand filtration.

(2) Sand filtration may be either,—
    (a) Natural.
    (b) Artificial or slow.
    (c) Mechanical or rapid, without coagulants.
    (d) Mechanical or rapid, with coagulants.

(3) Complete artificial systems on the best European model cost exclusive of land about $70,000 per acre, and filter 2,000,000 gallons per acre in each twenty-four hours.

(4) Complete mechanical systems on the best American model cost exclusive of land about $20,000 for a capacity capable of filtering 2,000,000 gallons in each twenty-four hours.

(5) It is a fact, established beyond dispute, that the filtration of contaminated waters will reduce the disease and death-rates from typhoid fever, cholera, and like diseases.

(6) Cities supplying plenty of clean and wholesome water attract population and every sort of manufacturing enterprise.

(7) Tests should be made in each locality to determine which system of filtration is best adapted to the existing local conditions.

4

PURIFICATION OF CITY WATER-SUPPLIES
BY SAND-FILTRATION.

By FRANK J. FIRTH, Philadelphia.

REPRINTED FROM
"THE ANNALS OF HYGIENE,"
JUNE, 1896.

Purification of City Water-Supplies by Sand-Filtration.

BY FRANK J. FIRTH, PHILADELPHIA.

ORIGIN OF SAND-FILTRATION.

THE tendency of men to congregate in cities and towns located on the water-courses of the country, whether navigable or not, brings with it a demand for important supplies of clean and wholesome water for public use, while at the same time it causes a pollution of the streams that are naturally relied upon to furnish these supplies. Having polluted the waters, destroying both their cleanliness and their wholesomeness, there comes at once a demand for purification. The purification of water by artificial methods involves a large outlay of money, which must be contributed directly or indirectly by the residents of the city or town to be supplied. Long discussions and delays precede definite action, and it not infrequently happens that some terrible epidemic of water-borne disease, such as cholera or typhoid fever, operating upon the fears of the people is needed to end discussions and delays, substituting therefor definite and intelligent action.

The ideal public water-supply is drawn from a mountain-stream or lake, that is and will always continue to be uncontaminated. Few places are so located as to make it worth while to waste much time in seeking such an ideal supply. It is either a myth or it is a practical impossibility because of financial conditions. It is wise to recognize the impossible where it clearly exists, and if the best is not to be had endeavor to attain as nearly to it as may be practicable.

Storage Reservoirs.—It is usually wise and economical in arranging any city water-system to provide sufficient storage reservoir capacity to prevent any interruption in the daily supply owing to accidents. A river flood or a broken pump should not be allowed, under any circumstances, to deprive the people of an article so essential to their well-being as is a supply of clean water. Storage reservoirs, when properly constructed and cared for, may be relied upon, incidentally, to effect a marked improvement in the quality of the water. Under favorable conditions they should yield what one of their advocates recently described as "fairly good water."

As compared with a polluted raw water it is undoubtedly a fact that sedimentation under favorable conditions may yield a "fairly good water."

It is by no means a fact, however, that such a water should be accepted by any community as a satisfactory answer to public demand for a clean and wholesome water. The only method at present known, and extensively used, for purifying contaminated waters and fitting them for safe use is sand-filtration. It is the purpose of this article to present in an elementary way some of the facts as to the development of the purification of water by sand-filtration. It is not the intention to attempt a scientific treatise on bacteria, nor to enter upon a discussion of the many problems in civil or mechanical engineering involved in the construction and operation of filter plants. The intention is rather to state in a plain way a few of the facts that every citizen may and should know before taking sides for or against this method of purifying the water needed for his daily use in home or factory.

All Filtration is Sand-Filtration.—In the first place, it may remove some misunderstanding and many doubts to point out that all filtration, worthy of the name, is sand-filtration. The sand may be from the sea-shore, from wells, or natural banks, or it may be an artificial product resulting from crushing certain varieties of quartz. Whatever may be its source, it is nearly enough correct for our present purpose to call it all sand and to say that all filtration of city water-supplies is sand-filtration. There are various methods of sand-

filtration. It may aid in their comprehension if they are classified and described as,—

(1) Natural sand-filtration.

(2) Artificial sand-filtration.

(3) Mechanical sand-filtration,—(a) without coagulants; (b) with coagulants.*

Natural Sand-Filtration.—The term "natural sand-filtration" is popularly used in this country to describe what is really "artificial sand-filtration." No scientist at home or abroad would voluntarily select the term "natural sand-filtration" to describe a process that is, at best, merely an artificial effort to imitate nature. It will, perhaps, be correct to say that the only filtration properly designated "natural sand-filtration" is filtration through sand in its original place, such as may be found in the case of the spring water as it comes bubbling through the earth clear, cool, and wholesome,—frequently sterile as the scientists say. Where springs do not exist, wells are driven, and the water drawn from their depths has been subjected to "natural sand-filtration." Small communities, unable to secure satisfactory supplies from springs or ordinary wells, sometimes resort to Artesian wells or to wells or channels located

* Mr. John C. Trautwine, Chief of the Bureau of Water, Philadelphia, recently suggested that instead of the misleading terms "natural sand-filtration" and "mechanical sand-filtration," now in popular use, the terms "slow sand-filtration" and "rapid sand-filtration" be adopted.

7

adjacent to rivers or other water-courses, securing a "natural sand-filtration" through the body of sand or earthy material lying between the river-bed and the collecting wells or channels. All of these methods of filtering water contemplate the passage of the water through undisturbed natural beds of sand or earthy materials, and they are properly described as representing "natural sand-filtration."

Artificial Sand-Filtration.—It is quite evident these methods are not available when the demands of large communities must be provided for. When these needs arise, the first resort is naturally to what may be designated as "artificial sand-filtra-tion." This method is a mere imitation on a large scale of nature's process, as it is popularly called. With the well idea in mind, large excavations are made in the ground to be used as shallow reser-voirs and filter-beds. The bed and sides of these reservoirs are made tight either by clay puddling or by the use of masonry or both. Under-drains are provided to carry off the filtered water, and the shallow basin or reservoir is filled to a depth of three to five feet with gravel and sand. Raw wa-ter (that is water as it comes from the river, pond, lake, or other natural source) is then allowed to flow into the reservoir over the sand. It gradually strains through the sand leaving the dirt, vegetable matter, and other impurities on the surface of the sand, while the clean-filtered water settles into the under-drains and is carried off to the pump-house or reservoir for delivery to the consumers. This, in a brief and elementary way, represents the simplest form of what is improperly called "natural sand-filtration," but properly "artificial sand-filtration."

There are many interesting details entering into the construction and operation of such a filter, but they are scientific and engineering details, and as such are foreign to the purpose of this article. It is quite evident that as the water passes through the bed of sand depositing its impurities upon the surface (and 80 or 90 per cent. are deposited literally on the surface, say one inch, and do not penetrate the lower layers) the surface of the sand will gradually become coated so as to interfere with and finally to stop the flow of the water through the sand bed. This is exactly what hap-pens and makes it necessary to draw off the water in order to clean the surface of the bed by hand-scraping. Of course, the filter-bed is out of use during the time it is being thus cleansed.

The first developments in "artificial sand-filtra-tion" are thus suggested. There must be a storage reservoir of sufficient capacity to supply the city with filtered water while the beds are being cleansed, or there must be a sufficient number of surplus beds to insure an area for constant use equivalent in capacity to the full demand upon them, and allowing whatever proportion may be needed to be out of service undergoing cleansing. The area of these surplus beds must, of course, depend on the character of the water to be filtered. If the raw water is heavy with sedimentary matter, it is evident the beds will be choked and require scraping much more frequently than will be the case if the raw water is reasonably clean. It is a serious matter to have filter-beds rapidly placed out of service owing to heavy deposits of sediment in their surfaces. It involves a large outlay for surplus beds. It costs largely for cleansing and waste or replacement of sand. It risks the beds being all out of service under exceptional flood conditions, thus forcing a resort to the raw water as the direct supply to the city while the beds are temporarily out of use.

These difficulties have been partially met in practice by the next development in "artificial sand-filtration." Large sedimentary basins are interposed between the river or other source from which raw water is drawn and the filter-beds. By allowing the raw water to stand in these basins twenty-four hours or more a large quantity of the heavier impurities settles, and the water passing onto the filter-beds is cleaner, and the life of the beds between the necessary scraping periods is advantageously lengthened. These sedimentary basins may be four in number, so that one may be receiving raw water; another standing to allow it to settle; another discharging the settled water on

the filter-bed, and the last out of service being cleansed of its sedimentary deposit. It is evident the addition of the sedimentary basins, while of great value, materially increases the investment cost of the plant.

The "artificial sand-filter," with its sedimentary basins and carefully-constructed sand-beds, may be located in a section where for days, weeks, or months, during the winter, there is freezing weather. The filter-basins are shallow ponds of almost still water and they readily freeze, just as would any shallow mill-pond, without requiring exceptionally low temperature. Water freezes on minimum temperature, and not, as some writers assume, on average temperatures. A few zero days in a month may not seriously affect an average temperature for the month, but they are sufficient to produce solid ice and to materially interfere with the successful operation of an "artificial filter-bed." This condition of affairs brings with it the next development. Filter-beds in freezing localities are, or should be, covered to protect them against the weather. The covering is of various forms, but usually, in the best construction, of masonry arches covered with earth. It is desirable to cover "artificial sand-beds" for other reasons than because of freezing weather. It will be remembered that it is necessary to cleanse the beds by scraping the sediment from their surfaces, and that until this is done they must remain out of service

9

when once clogged. This scraping can only be done satisfactorily when the surface of the bed is dry enough for laborers to work upon it. It is evident it would not be convenient or economical, even though possible, to cleanse the bed during a period of rain or snow.

The complete "artificial sand-filtration" plant must, therefore, consist of sedimentary basins; filter-beds with an ample surplus area for cleansing; covers against frost, rain, and snow. Such complete beds cost about $70,000 to $75,000 per acre to construct, exclusive of the value of the land, and their capacity under the best European practice is about 2,000,000 gallons of water per acre in each twenty-four hours.

There are continuous and intermittent "artificial sand-filters," each having certain relative merits, but it is not worth while to refer to them in any popular article. They may be dismissed as engineering and scientific details of operation rather than construction.

Mechanical Sand-Filtration—without Coagulants. —The next development in the purification of water by sand-filtration is found in what are known as the "mechanical sand-filtration" processes. Development may not, in its early stages, yield the desired improvement. It is a consequence of recognized imperfection or defect in existing conditions. The imperfections in artificial sand-filtration are:

(1) It requires a large area of level land not always to be had.

(2) It is a costly system to construct.

(3) It is inconvenient in cleansing and replacing the sand.

(4) It is slow in operation.

These recognized imperfections call for an effort to overcome them if possible, taking care, however, not to sacrifice in this effort any of the beneficial results attained by the artificial sand or slow filtration method.

There are various practical reasons for the existence of the mechanical sand types of filtration. It has been stated early in this article that cities and towns approach the filtration of their water-supplies through many discussions and delays. It is not so with the individual resident. The manufacturer must have clean water at once, and to do this he must provide his own filter plant. He can not construct an "artificial sand-filter," with its large acreage and slow work. He must have its substantial equivalent in some less cumbrous form. Instead of the shallow reservoir he erects a wooden or iron tank. He places in the bottom of this tank the same kind of a bed of sand that would be used in an artificial sand-filter, through which the raw water entering his factory is to be strained. He connects the tank with a city-supply or his own pumps, provides proper outlet-pipes for the clean water, and proceeds to operate his filter.

He soon finds the cleansing of the surface of his sand-bed by scraping it too troublesome and costly. Every need in life suggests to some inventive mind its solution. Various methods of cleansing the sand-bed are proposed and tried. The one best adapted to the need is ultimately evolved, and to-day, instead of the manufacturer scraping the surface of the sand in his tank by hand labor, he operates a series of valves obtaining a reverse current of water through the bed, and, if he so elects, aids the work by the use of mechanism driving rakes or stirrers through the sand while the washing is progressing. In this way the dirtiest sand-bed may be thoroughly cleansed and ready for service again in about ten minutes.

Here we have then "mechanical or rapid sand-filtration" without the use of coagulants, and intended, chiefly in this stage of the work, to yield a clean water fit for fine manufacturing and without any reference to the removal of bacteria or disease-germs. Clean water and not wholesome water is the primary object of the manufacturer, and he is, therefore, able to filter at much higher rates per acre for twenty-four hours than would be thought of in an "artificial sand-filter." While 2,000,000 gallons per twenty-four hours per acre is the limit where "artificial sand-filtration" is resorted to for the purification of city supplies, the manufacturer obtains his clean water by the use of "mechanical sand-filtration"—without the use of coagulants—at the rate of 150,000,000 gallons, or more, if desired, in twenty-four hours per acre of sand surface.

It should be remembered that the term "mechanical" attaches to this type of sand-filter only because of the mechanism used to cleanse the sand-bed by the resort to reserve currents of water with or without power-rakes to stir the sand while being cleaned. Precisely similar methods have been tried with the "artificial sand-beds" where mechanism has been used to stir the surface of the sand while a stream of water was thrown across it to carry off the sedimentary deposit in this way instead of by hand labor scraping. The method is not an economic success in the "artificial sand-beds" because of the large area of the beds. What may be accomplished in a tank of fifteen feet in diameter may not so readily or efficiently be used on a half acre, or acre and a half, "artificial sand-bed." "Mechanical sand-filters—without coagulants"—are made in many forms and for many uses, including the cleansing of city water-supplies. While the earlier forms of mechanical filters were erected solely for the purpose of cleansing water, and before the relation of certain forms of bacteria to the wholesomeness of water was at all well understood, it is nevertheless claimed that these filters effect a marked improvement in the wholesomeness as well as the cleanly appearance of the water.

11

Mechanical Sand-Filtration – with Coagulants.—
The next development appears to be found in the
process of "mechanical sand-filtration—with coagulants." The tank construction and general mechanism for city use are substantially the same in
"mechanical sand-filters" whether used with or
without coagulants. The use of a coagulant certainly represents a stage in the process of filtration development. It did not originate in connection with the use of "mechanical sand-filters," but
was a natural and scientific result of an effort to
counteract the evil effects of the deleterious matter
found in contaminated waters. The coagulants
commonly made use of are metallic iron, commercial alum, or sulphate of aluminum, and in a
smaller and rather experimental way, certain other
salts.

The use of alum as a coagulant is so common as
to be practically universal in the most effective
forms of mechanical sand-filters. Notwithstanding
its extensive use for many years in small filters
for household use; in the much larger filters used
in large office-buildings, hotels, hospitals, etc., in
almost every variety of factory requiring clean
water, and in filtering the public water-supplies of
a hundred or more cities and towns, the knowledge of the way in which the alum acts and the
exact results obtained by its use are, it is safe to say,
imperfectly understood by the general public. It is
known to scientists that the sulphate of aluminum
(commercial alum) when used in moderate quantities, say half a grain to a gallon of raw water containing lime, magnesia, etc., is decomposed into
sulphuric acid and aluminum. The acid combines
with lime or other suitable material that exists in
or may be added to the water, while the hydrate of
aluminum is formed and precipitated as a gelatin-
our mass on the surface of the sand-bed, carrying
with it certain of the organic and other impurities
in the water. With ordinary care in operation,
such as must be exercised in the use of any mechanism, no more alum need be placed in the raw
water than will be in this way decomposed and
made useful. The resulting compounds following this decomposition are deposited with other
matter on the surface of the sand and are washed
away with all sedimentary matter accumulated
on this surface at the time of the cleansing of the
sand-bed. The gelatinous hydrate of aluminum is
said to perform a useful function in partially filling the small interstices between the grains of
sand, thus making it possible to stop the coagulated small germs or bacteria on the surface of
the bed even when filtration is progressing at the
extraordinary rate of 100,000,000 gallons or more
per acre per twenty-four hours. Proposals submitted to the city of Philadelphia indicate that
complete mechanical sand-filters, according to the
best American methods, can be erected at a cost,
exclusive of the value of the land, of about $20,-

000, for a capacity capable of filtering 2,000,000 gallons in each twenty-four hours. This result is so amazing as to cause it to be viewed with a natural distrust by the scientists whose life-long experience has been with the "artificial or slow sand-filters;" they have found it necessary, from various causes, to limit to say a rate of 2,000,000 gallons per acre per twenty-four hours. While this distrust is natural and proper, it should not lead any disinterested investigator to condemn unheard a process of filtration which, if effective, offers many practical and economic advantages as compared with "artificial sand-filters" of Europe.

Louisville, Kentucky.—The mechanical sand-filters using coagulants should be subjected, in actual operation under city conditions, to a rigid scientific scrutiny, as is now being done at Louisville, Kentucky, which to-day stands in advance of every American city in its method of studying its own local water problem. It has not yet made known the result of its investigation, but when this is done it is safe to say it will be the most valuable contribution thus far made to accurate knowledge upon the subject of mechanical sand-filtration. Thus far the reliable bacteriological analysis of the filtrates from the mechanical sand-filters have been comparatively few in number and almost entirely confined to laboratory experiments or experiments on a laboratory scale. The results attained under these favorable conditions are by no means conclusive, but they certainly justify such intelligent investigation under city conditions as that now being made in Louisville.

Filters in the United States.—It is something in favor of the mechanical sand-filters that our own people, who are quick to appropriate good, economic methods of every sort, have with practical unanimity adopted the mechanical sand system in preference to the artificial sand system on the European model. While hundreds of cities and towns in the United States and Canada filter their public water-supplies by mechanical sand methods, there is but one artificial sand-filter in this country (at Lawrence, Mass.) that is recognized as worthy of note. The Lawrence plant is a comparatively small one, filtering the water-supply of a city of but 50,000 inhabitants, and it is by no means a complete filter of its type. It has no sedimentary basins between it and the Merrimac River, from which it is supplied. It has but a single bed without divisions, thus making it difficult to cleanse it satisfactorily. It has not a tight masonry bottom, and, in consequence, it allows the river and surface-waters to enter the effluent in a small way and thus pollute it. It is not covered as a protection against frost, rain, and snow. Notwithstanding these defects, due to an enforced economy in its construction,

13

this Lawrence plant has demonstrated its practical value in greatly diminishing the disease and death-rates from typhoid and like causes.

Results of Filtration.—Any properly-constructed and operated artificial or mechanical sand-filter will remove from water the sedimentary matter visible to the ordinary sight, yielding a clean, clear filtrate in place of a water clouded and offensive owing to its containing mud, coal-dust, or worse impurities. The clean, filtered water thus obtained is certainly improved for ordinary household uses as well as made suitable for practically every variety of manufacturing industry. As to this there is no question or doubt. If this was all that could be accomplished by filtration, its use would clearly justify its cost and would pay a large return thereon in attracting population and manufacturing industries to the city where a plentiful supply of clean water was an assured fact.

But filtration does much more than this. It not only cleans foul and contaminated waters, making them more agreeable to the sight and suitable for many uses; it is an established fact, as to which all well-informed authorities are agreed, that the filtration of city water-supplies gives a wholesome in place of a disease-bearing water, with an unmistakable diminution in the number of cases and deaths from such water-borne diseases as cholera,

typhoid fever, etc. The experience of Berlin, Hamburg, Altoona, Lawrence (Mass.), and many other places is a matter of authoritative public record and questioned by no one familiar with the facts. It will not be forgotten that all varieties of filters are artificial creations, machines needing intelligent care. It is possible with filters, as with all other types of mechanism, to meet with accidents in operation. With intelligent care such accidents should be few in number and unimportant in result, but it is a fact accidents may occur. In such event the worst result would be that water might have to be temporarily drawn direct from the river or other source without passing through the filter-bed, and the public supply during a few days or hours might then possibly be as bad as it would certainly be at all times if the filter-bed did not exist. To condemn the use of filter-beds because they may at rare intervals and for short periods be subject to interruption or diminution in efficiency from unforeseen accidents would be about as reasonable as it would be to discard pumping-engines from a like cause.

Conclusions.—In conclusion, the following facts should be known and always remembered:
(1) All filtration is sand filtration.
(2) Sand filtration may be either,—
    (a) Natural.

11

(b) Artificial or slow.

(c) Mechanical or rapid, without coagulants.

(d) Mechanical or rapid, with coagulants.

(3) Complete artificial systems on the best European model cost exclusive of land about $70,000 per acre, and filter 2,000,000 gallons per acre in each twenty-four hours.

(4) Complete mechanical systems on the best American model cost exclusive of land about $20,000 for a capacity capable of filtering 2,000,000 gallons in each twenty-four hours.

(5) It is a fact, established beyond dispute, that the filtration of contaminated waters will reduce the disease and death-rates from typhoid fever, cholera, and like diseases.

(6) Cities supplying plenty of clean and wholesome water attract population and every sort of manufacturing enterprise.

(7) Tests should be made in each locality to determine which system of filtration is best adapted to the existing local conditions.*

---

* The writer of this article is a firm believer in the purification of public water supplies by filtration. He does not favor any particular mechanical or other system, whether slow or rapid, but urges each community to determine by impartial scientific investigation which of the various systems of filtration is best adapted to its own local use.　F. J. F.

Louisville Experiments.

HE Louisville Water Company, having determined to purify their entire water supply, decided in the spring of 1895 to institute a series of practical tests on the various systems of mechanical filtration that might wish to participate therein. The reason for limiting the experiments to filters of the mechanical or rapid type was probably due to the very large amount of suspended and earthy impurities carried by the water, which would necessitate too frequent cleansings of any filter-bed system.

We erected our filter during the month of June, 1895, but it was not until November of that year that the balance of the filters were completed and the official test begun. The experiments were concluded on the 1st of August, 1896, being carried through all the worst stages and conditions of the river. The experiments were conducted at the pumping station, the water supplied to the various filters being taken directly from the river. Each filter was comfortably housed in separate buildings erected by the Water Company for this purpose, while the laboratory occupied a spacious building of its own. Mr. George W. Fuller, the eminent bacteriologist, well known in connection with the researches of the Massachusetts State Board of Health, was in charge, with Mr. Robert Spur Weston as assistant, and the laboratory was manned in all by from six to nine competent experts. The work was divided into several divisions, and the tests frequently conducted night and day for several weeks at a time, especially when the river was in an extraordinary turbid condition; and it may be said in reference to these tests that they were conducted on the most thorough and scientific basis, nothing being spared in order to get at the true value of filtration; and when these results are published, this contribution to the literature of filtration will have surpassed anything heretofore made, and will be of inestimable value to all who are considering the best system for purifying their water supply. These reports will not be made public until they appear in the regular annual report of the directors of the Louisville Water Company, which will probably not be ready for distribution before March, 1897; and until then we are not at liberty to make known any of the results accomplished by our filter during these tests; but in lieu thereof, have reprinted several brief extracts from the tests made by the City of Providence, R. I., which are of the same general character; and sufficient to convince anyone that our system of mechanical filtration far surpasses the old European style, or the filter-bed system, and establishes beyond dispute, the high plane of mechanical filtration and the superiority of the Jewell filters.

16

Bacteriologic Results from Mechanical Filtration.

[Abstracts from a Report made by E. B. Weston, C. E., of ·Providence, R. I., upon Experiments made by Natural Filtration and in the use of the Jewell Filter for Period of Nine Months.]

Read at Meeting of the American Public Health Association, Denver, Colo., October, 1895.

BY GARDNER T. SWARTS, M. D.
SECRETARY OF STATE BOARD OF HEALTH, RHODE ISLAND,
PROVIDENCE, R. I.

Reprinted from the Journal of the American Medical Association, January 18th, 1896.

Bacteriologic Results from Mechanical Filtration.

By GARDNER T. SWARTS, M. D.

T the last meeting of this association at Montreal the statement was made in the report of the Committee on Water Supplies that no data had been available to show that filtration by the so-called mechanical methods was successful in removing bacteria. The writer at that time referred to experiments which had been made in the City of Providence, R. I., in order to determine this question for the purpose of establishing a plant capable of filtering 15,000,000 gallons daily if the experiments were successful.

The figures showing these results were not at that time available, and as they have never been published and as no experiments of a similar character have been made, it seems desirable to place these facts before the Association, inasmuch as many municipalities are agitated over the advisability of introducing the so-called natural or sand-bed filtration or mechanical filtration.

The mechanical form of filter used in the experiments was of the type in which quartz or sand is used as a supporting bed to a film of precipitated coagulant or fixative of organic matter, produced by the introduction into the water,

before filtering, of some chemical such as iron or alum; a filter which is also cleansed by means of a reversed current of the water passed through the filter assisted by the use of a rake made to revolve in the bed of the quartz while the washing is being done.

The filters used in this line of experiments were two of the natural sand-bed form imitating the usual filter bed. The mechanical form was represented by one of the New York Filter Company's filters and one of the Jewell filters. After the first seven months the sand filters were discontinued, it having been satisfactorily ascertained that the length of run was much less than the mechanical filter before the bed became clogged and the rate of flow in the natural bed was but 3,000,000 gallons per acre in twenty-four hours, while the mechanical filter was run at the rate of 128,000,000 gallons per acre in twenty-four hours. The efficiency of removal of bacteria was not as high, and the results variable, either as the result of cracks in the filter or from some unknown reason. Although both of these natural filtration beds were constructed exactly alike, the results from the second were much poorer than the first. When the natural bed was transformed or assisted by the addition of alum, thus converting it into a mechanical filter, the removal of bacteria was increased to nearly the same as on the Jewell filter, but the length of the run was correspondingly decreased.

The sand used in the natural beds was a natural river sand, not over sharp, while the sand used in the mechanical filter was crushed quartz having sharp edges.

In the beginning of the experiments the New York filter gave such varied and unreliable results that its use was abandoned, while the Jewell filter was continued in use during the whole series of experiments, which lasted for a period of about ten months, the working of the mechanical parts of the filter being perfectly satisfactory and the results obtained being successful.

The filter bed used in this mechanical filter was two feet ten inches in depth, supported upon a base of iron with circular perforations of about four inches in size, which were covered with screens. The crushed quartz used was the "effective size" of 0.50 millimeters. The filter was washed by a reverse current which caused the quartz to boil. The agitation and friction of the particles were increased by means of a rake with long teeth which revolved about a central column in the filter; the rake penetrating the bed by a screw motion from top to bottom.

From the various kinds of congulant or precipitant used basic sulphate of alumina was selected as being the most satisfactory and effective and was used in all the experiments mentioned. The amount of alumina used was ½ grain to the gallon of water filtered, a lesser quantity failing to satisfactorily remove the organisms, while the amount of ¾ or 1 grain per gallon did not increase the removal of the bacteria, while the efficiency of the filter was greatly decreased by reducing the amount of the flow through the filter bed.

The alumina was applied in a free flow at the beginning of a run by pouring into the filter, as the water entered, a pint of the congulant containing about 911 grains of sulphate of alumina for an average flow of 128,000,000 gallons per acre. The solution was made by adding one part of the alumina to six parts of water; as a result of this addition there forms a white flocculent precipitate over the surface of the grains of quartz and is the actual medium through which the filtration takes place, the quartz serving merely as a supporting bed or sieve. It required about six minutes to form this layer. When applied at the rate of a drop at a time and not in a "free flow" it required about half an hour before the filtering layer would be formed. As soon as the filtering layer was formed the alum solution was dropped in continuously during the run from a regular stop at the rate of a drop a second. The effect of the presence of this layer was to reduce the head or pressure .28 of a foot for 128,000,000 gallons per acre. The depth of the water above the bed at the commencement of the run was nine inches; the average length of the run was about eighteen hours.

Under these conditions it was determined how long after the commencement of the run the filtering ability was at a maximum and also the capacity of the filtering media to remove organisms and also the possibility of removing organisms foreign to river water and simulating pathogenic bacteria in their life history. In this last experiment the Cruikshank bacillus and the bacillus prodigiosus were used, since from their pathogenic properties they could be readily distinguished from the water bacteria.

For an understanding of the proportion of bacteria found in the applied water and the number to be found in the filter water, table No. 3 of the report is here appended.

As a result of the whole series of experiments the totals shown in table No. 3 will give an idea of the averages. In consideration of this table it must be remembered that the introduction of only one result, which may be far below the average, will readily reduce what would otherwise be a most favorable average, to a lower point. This one result might occur from a temporary contamination of the effluent pipes at the time collecting the sample, and which might not represent the exact results of filtration.

During the application of the cultures of bacillus prodigiosus in large quantities suspended in the nutrient media, the numbers of the common water bacteria materially increased in the effluent, the particles of quartz becoming covered with a slimy brownish deposit. Unsuccessful efforts were made to cleanse the quartz of this growth by steaming and boiling the quartz for one hour. Finally on the application of a solution of one pint of caustic soda to twenty-four parts of water and steaming, the normal white color of the quartz returned. The efficiency of the filter was raised by this process of cleansing from 92.8 per cent. to 98.8 per cent. As to the mooted dangers attending the use of alum in the applied water and which is held up as a warning by the opponents of mechanical filtration, this much may be said in reference to this series of experiments:

While it was necessary to add half a grain of sulphate of alumina per gallon of water filtered in order to obtain the most satisfactory results, yet upon comparison by the most careful chemical tests of the water applied to the filter and that of the effluent, there was found to be less alum in the filtered water than in the river water itself.

Inquiry from numerous manufacturers using alum as a precipitant in various quantities in excess of the amount used in the experiments, revealed in no instance any incrustation or scaling in the boilers where such filtered water had been used. Communication with the various boiler insurance companies elicited no report of scaling where such water was used. There is no recorded instance where alum-treated water as a drinking water has produced any ill effects upon the consumers.

This work was done by order of the City Council of the City of Providence and under the direction of a commission consisting of the Superintendent of Health, the City Engineer and the Commissioner of Public Works. The immediate supervision of the operation was under the supervision of Dr. C. V. Chapin, the Superintendent of Health and a member of this Association, while the application of the various tests was made under the direction of Mr. Edmund B. Weston, C E., from whose computations and reports these abstracts have been taken. Most of the bacteriologic work was done by the writer.

Inasmuch as the writer, as well as every person connected with the experiments, commenced the investigation with the firm belief that successful mechanical filtration was not possible from a bacteriologic view, it must be stated now, after examination of these figures, that mechanical filtration under these conditions can be firmly indorsed.

The foregoing paper and table of percentages was briefly discussed, first by Dr. P. H. Bryce, who thought the facts contained therein were most valuable and that he felt personally that the thanks of the Association were due to Dr. Swartz for submitting them, and by Rudolf Herring M. A. S. C. E.

TABLE NO. 3.—FILTRATION EXPERIMENTS WITH JEWELL FILTER.

Growth of about ninety hours, of water bacteria, in the sample of applied and filtered water which were taken at the same hour; which was one hour or more after the water commenced to flow from the filter.

Date.	Gallons of Water Filtered per acre per Twenty-four hours.	Bacteria per Cubic Centimeter. In Applied Water.	Bacteria per Cubic Centimeter. In Filtered Water.	Per Cent. of the Applied Bacteria Removed.	Average Percentage of the Applied Bacteria Removed.	Grains of Sulphate of Alumina added.
1893.						
July.						
20	125,000,000	2,000	11	99.5		0.75
21	122,000,000	2,477	16	99.8		0.90
Oct.						
3	125,000,000	905	6	99.3		0.60
4	129,000,000	610	2	99.7		0.58
5	131,000,000	4,002	25	99.4	99.5	0.55
17	125,000,000	6,175	26	99.6	(By totals, 99.6)	0.57
27	122,000,000	6,700	41	99.6		0.61
30	128,000,000	1,700	7	99.6		0.58
31	131,000,000	400	9	97.8		0.50
Nov.						
1	132,000,000	15,112	19	99.9		0.61
2	123,000,000	6,950	26	99.6		0.83
3	122,000,000	9,400	50	99.5		0.84
4	122,000,000	3,400	63	98.1		1.20
9	125,000,000	2,900	26	98.8	99.2	0.60
11	125,000,000	3,650	25	99.3	(By totals, 99.5)	0.82

COMMENCED TO USE THE BACILLUS PRODIGIOSUS.

Date.						
Nov.						
23	120,000,000	15,950	218	98.6		0.60
24	132,000,000	7,600	384	95.2		0.50
Dec.						
2	125,000,000	4,900	100	98.1		0.75
4	125,000,000	4,475	91	98.0		0.60
1894.						
Jan.						
2	132,000,000	2,150	94	95.6		0.85
3	127,000,000	2,000	118	94.1		0.84
4	134,000,000	2,275	44	98.1		0.85
5	130,000,000	1,025	80	98.9	98.1	0.82
6	130,000,000	2,375	184	92.3	(By totals, 96.9)	0.58

CEASED TO USE BACILLUS PRODIGIOSUS.

Date.	Gallons of Water Filtered per acre per Twenty-four hours.	Bacteria per Cubic Centimeter. In Applied Water.	Bacteria per Cubic Centimeter. In Filtered Water.	Per Cent. of the Applied Bacteria Removed.	Average Percentage of the Applied Bacteria Removed.	Grains of Sulphate of Alumina added.
1894.						
Jan.						
9	130,000,000	1,850	54	97.1		0.60
10	134,000,000	800	24	96.8		0.84
11	130,000,000	780	20	97.3		0.61
12	132,000,000	350	52	85.1		0.81
13	132,000,000	600	38	94.0		0.72
15	134,000,000	925	46	90.5		0.74
16	134,000,000	375	44	88.3		0.58
17	130,000,000	2,150	64	97.0		0.82
18	134,000,000	1,500	62	95.9		0.54
19	136,000,000	1,450	80	94.5		0.83
20	130,000,000	2,800	58	97.9		0.72
22	132,000,000	3,350	62	98.1	94.8	0.85
23	132,000,000	2,300	64	97.2	(By totals, 96.3)	0.80

WASHED FILTER BED WITH CAUSTIC SODA.

Date.						
Jan.						
24	128,000,000	2,100	6	99.7		0.60
25	125,000,000	2,325	18	99.2		0.82
26	128,000,000	4,850	54	98.8		0.58
27	129,000,000	4,875	72	98.5		0.58
29	126,000,000	1,575	83	94.8	98.2	0.50
30	130,000,000	1,400	28	98.8	(By totals, 98.5)	0.58

Extracts from Report of the Results obtained with Experimental Filters.

By EDMUND B. WESTON.

THE following is a brief summary of the results obtained with the Jewell Mechanical Filter while filtering at the rate of 128,000,000 gallons per acre per twenty-four hours, when six-tenths (0.6) of a grain of sulphate of alumina was being used per gallon of water:—

	Per cent.
Water bacteria	98.7
Applied "Bacillus Prodigiosus" removed	99.8
Albuminoid ammonia	70.0
Ready-formed ammonia	91.0
Color removed during the day	78.0
Color removed during the night	66.0

THE BACTERIOLOGICAL PURIFICATION OF THE WATER.

Including samples taken thirty minutes or less after water commenced to flow from filter, and those taken at the same hour as the APPLIED WATER (which was one hour or more after water commenced to flow from the filter).

	Percentage of Removal.
End Growths	98.6
Growths of about 90 hours	98.7

Including samples taken thirty minutes or less and all samples taken one hour or more after the water commenced to flow from the filter.

	Percentage of Removal.
End Growths	98.6
Growths of 85 hours or more and End Growths	98.7

"This table shows that the average efficiency of the filter for removing water bacteria was 98.6 per cent."

THE CHEMICAL PURIFICATION OF THE WATER.

"Table No. 20 shows, from an average of three analyses, a reduction of Albuminoid Ammonia by Filtration, of 70.0 per cent., and a reduction of Ready-formed Ammonia of 91.0 per cent."

THE REMOVAL OF BACILLI PRODIGIOSI.

"Table No. 19 shows that the average percentage of the Applied Bacillus Prodigiosus that was removed from the water by filtration was 99.8 per cent."

THE WASHING OF THE FILTER BED.

"Our experiments have shown, I think, that the filter bed could be very thoroughly washed by the aid of the mechanical rake or agitator.

"From a mechanical standpoint, the working of the Experimental Jewell Mechanical Filter, throughout the experiments, was very satisfactory."

COAGULANTS.

FROM CITY DOCUMENT No. 15. THE CITY OF PROVIDENCE, R. I.

Report of the Joint Special Committee to Examine and Report Relative to the Pollution of the Water Supply and the Best Method of Filtration.

"ANY persons who are unacquainted with the results actually obtained are prejudiced against the use of alum, or aluminum sulphate in any form, in the process of filtration. Your committee would refer to the elaborate discussion of this subject in the Seventh Annual Report of the Superintendent of Health of this city, in which he shows that the amount of alum used is absolutely harmless, even if retained in the filtered water, while he refers to the experiments which prove that, although alum was found in the unfiltered water, there was no trace of it in the effluent from filters in which this coagulant was used.

"Many of our best water supplies contain more alumina than is used in any of the better of the mechanical systems in treating water similar to that of the Pawtuxet. Moreover, as above stated, if the coagulant is supplied in the proper manner no trace of it will be found in the effluent. Your committee itself witnessed the most searching test of this known to chemists, the well known color test with logwood, by which not the slightest

trace of the alum could be discovered in the effluent from a filter using more alum than would be employed in filtering the Pawtuxet.

"The wide diffusion of aluminiferous earth (red clay) in almost all soils renders it highly probable that this is one of nature's agents in producing the sparkling brilliancy of our natural spring waters.

"Your committee is satisfied that the present scientific use of alum, or, better still, of basic aluminum sulphate, as a continuous coagulant, is one of the greatest aids to an efficient result in mechanical filtration, and that it can be used with absolute certainty of no danger whatever resulting from it."

ACTION OF THE COAGULANT.

"If the diameter of matter floating about in the water is much less than that of the interstices between the grains of sand composing the filter-bed, such matter, except so much as is caught upon the sharp edges of the quartz, will go right through the filter-bed with the water.

"Now, if a substance could be introduced drop

by drop into the water before it comes to the filter-bed, which would have the effect of curdling this matter together, so that every one hundred or so of the smaller particles were made to join together and become one large particle, much as vapor or steam is condensed into drops, it would follow that they would be caught and held from going through the filter. This is accomplished by adding dissolved sulphate of alumina (alum) to the water as it flows to the filter.

"The amount required is from none at all to about three-quarters of a grain, according to the state of the water, say an average of from one-quarter to one-half a grain per gallon in the ordinary condition of the Pawtuxet. The introduction of such coagulant is regulated to the amount required in the simplest and easiest manner.

"This method of purifying water was known to the ancients. The action is the same as when coffee is clarified by means of the white of an egg. No white of the egg goes to the drinker of the coffee, it is all strained out with the grounds; and so no alum goes to the drinker of the water, it unites with the impurities in the water and settles in feathery flakes of insoluble aluminum hydrate on the top of a gravity filter, and is washed out with its accumulation of impurities when the bed is cleansed.

"The analysis of the purified water shows no trace of the alumina used, while the analysis of the wash water shows that the alumina is washed out with the other impurities. This feathery bed of precipitate flakes produced by the alum forms a filtering material of insoluble mineral matter which is well nigh perfect in its character. Bacteria are like the fine clay particles of some water, so small as to pass the sand or quartz, but they are caught by the feathery precipitate of aluminum hydrate, much as the bacteria contained in the air are prevented from entering a vial closed with sterilized cotton.

"Aluminum hydrate has also the remarkable property of taking up coloring matter. Whether this is wholly a physical or partly a chemical action is a mooted question. The yellow or brown stain or color that many waters have (due to dissolved vegetable matter, as is the case with tea or coffee, and is largely the case with the water of the Pawtuxet River) which remains, although the water has passed through the finest filter, is entirely eliminated by passing through a film of freshly precipitated aluminum hydrate. Some gelatinous vegetable impurities are also removed, and, as a consequence, water filtered through quartz, over which a film of the hydrate has been formed, is in a great measure freed from substances which yield what the chemists call albuminoid ammonia."

Analysis showing less Aluminum in Pawtuxet River Water after than before Filtration.

By PROF. JOHN H. APPLETON, of Brown University.

The larger figures signify parts (by weight) in one million parts of water (by weight). The smaller figures signify grains per American gallon of water (weighing 58,372.2 grains,

	Oxide of Iron. (Fe_2O_3).	Oxide of Aluminum, (Al_2O_3).
River Water......	.74	.57
	.013	.033
Filtered Water....	.20	.30
	.012	.017

—*Extracts from the Appendix to the Seventh Annual Report of the State Board of Health of Rhode Island.*

Coagulants Employed and Method of Using Them.

AMONG the many substances available for the coagulation or coalescence of the impurities of waters, basic aluminium sulphate and lime are principally, if not almost universally, used at the present time.

Lime has been used in various ways, chiefly for the reduction of hardness, and with the resultant secondary effect of clarifying and purifying organically the water thus softened. This re-agent has, of late, however, due to our efforts, come into extensive use as an auxiliary re-agent to aid in and insure the complete precipitation of sulphate of aluminium, especially in the treatment of soft or river waters during freshet seasons or heavy rises, upon which waters alumina compounds alone are of little practical value.

This combined method of treatment was originated by Frederick Arthur Paget and patented by him in England in the year 1874. As therein described, it consists of first treating the water with an alumina salt and subsequently with lime, the effect being an instantaneous and complete reaction between the two, whereby all of the alumina is thoroughly precipitated.

We have employed this method with most excellent results for several years and have found in all instances that not only were the waters greatly improved in quality, but that there was an actual saving of from 30 to 50 per cent. In the amount of aluminum sulphate used, at only a trifle expense for the lime.

Nearly all of our plants are fitted with complete apparatus for the use of both lime and alumina, either singly or together, as occasion demands.

The solutions of aluminum sulphate are prepared in moderately large cypress tanks, arranged in duplicate and connected to work alternately, so that ample time is allowed for the thorough preparation of solutions of any desired strength, and these tanks are constructed so as to operate automatically, thereby requiring very little attention.

The preparation of lime water is made continuous owing to its sparing solubility, and the tank for its use is fitted complete with all the necessary requisites, making them perfectly automatic, and about the only attention required is to maintain the supply of lime.

The method most generally employed for feeding these re-agents into the water is to use a small auxiliary pump, operated automatically, either by steam or mechanical connection, to the low service pump, supplying the filters. This arrangement insures both large and small pumps working with absolute certainty and uniformity, feeding a definite and constant quantity of the re-

27

agent to every gallon of the natural water. No more simple, effective, or positive device could be desired. Little or no attention whatever is required to keep it in perfect operation, and, moreover, it can be easily and readily adjusted to feed any requisite amount. Under this arrangement and with ordinary care, no trace of the coagulant will appear in the filtered water under any conditions.

On small plants or single filters, or those supplied by a head of water or not supplied by a separate low service pump, some slight modifications, dependent on local conditions, in the above described apparatus is necessary, as, for example, the use of a meter or water wheel to operate the auxiliary pump. The general plan, however, is the same, possessing all of the advantageous features above enumerated.

SPECIFICATION FOR ALUMINA

To enable those not fully conversant with the various sulphates of alumina on the market, we give the following percentages of the principal constituents:

A first-class product should contain—

	Per cent.
Alumina (Al_2O_3). Total	17 to $17\frac{1}{2}$
Alumina (Al_2O_3). Comb. as sulphate	$15\frac{1}{2}$ to $16\frac{1}{2}$
Alumina (Al_2O_3). Free (Basic)	1 to $1\frac{1}{2}$

and should not contain more than—

	Per cent.
Iron (calculated as Ferrous Sulphate)	$\frac{1}{4}$ to $\frac{1}{2}$
Insoluble and other foreign matter	$\frac{1}{2}$ to 1
Water	45 to 46

The above would insure the purchaser obtaining a product satisfactory for all requirements, and at a reasonable figure.

The limits we have given are not such as will bar out any reputable make of sulphate of aluminum, but do represent to the best of our ability and experience, an alum of moderate cost, and specially adapted for use in filters.

Twelve Foot Jewell Subsidence Gravity Filter.

Description of the Jewell Improved Gravity Filter, with Subsidence Basin.

UR standard gravity filters are all fourteen feet in height, varying in diameter and length according to the size of the plant and other local conditions. The tanks are generally built of well-seasoned select cypress or cedar, dressed on both sides to 2¾ inches, but are also built of iron or steel if desired. Both main and inner tanks are bound with extra heavy wrought iron bands, each provided with three pairs of improved draw lugs. The main staves are double grooved to receive the lower bottom and circular segments, which are also chimed and grooved around the inner tank. This method of construction insures absolutely no leakage between main and inner tanks, and affords an ample space between both tanks to carry off the wash or waste water. The timbers supporting the inner or filter bed tank are of hard wood and very large and strong, so that no weight or strain is carried on the staves. The upright timbers rest upon a heavy secondary floor laid upon the bottom of the main tank for the purpose of distributing the weight. Diametrically across the floor of the inner tank is laid the heavy cast iron manifold sections, which are held together with machine bolts. The strainer or branch pipes which are screwed into these manifolds at equal distances along same, extend to the side of the

inner tank, being cut in various lengths to fit the circle. These pipes are extra heavy wrought iron and capped on the ends by reducing elbows. The wash and collecting strainers are evenly spaced along these pipes and most securely fastened thereto by means of double brass clamps. The central manifold is provided with a flanged opening to which a central standpipe is bolted, and extends several inches above the level of the filtering bed. The upper bottom is cut out in the center to the area of this standpipe and a heavy flange is secured on the underneath by means of studs which pass through the bottom and enter the central manifold, thereby rigidly clamping all parts together. The stirring apparatus consists essentially of a heavy cross-arm, which swings on a vertical shaft extending from worm gear on top to the deflecting elbow in subsiding tank. Into this cross-arm are keyed three heavy horizontal shafts, which carry the swinging rakes or bars. These bars are secured in swivel collars which are held between two heavy set collars, having quadrant lugs which engage the swivel collars when the agitator bars are vertical in the bed. These rakes or bars are arranged to traverse the bed every three inches, thereby reaching all its parts. The worm and worm wheel are combined in one heavy cast iron casing, making a most substantial and perfect running mechanism. The casing provides for a well of oil in which the worm and wheel dip, so that both are continuously running in oil. The worm shaft is fitted with an end support secured to the cross beam on top of the filter, and is provided with either duplex friction clutches operated with one lever, or tight and loose pulleys with shipper rods, if preferred. The upper cross beams on which the worm casing and driving apparatus is mounted is hard wood and rigidly fastened to the main tank by heavy cast iron angle plates or brackets. The down draft pipe for carrying off the purified water extends from the bottom of one of the manifolds to near the floor of the subsiding basin, thence out through main tank to a cross in front of the filter. This cross is fitted with three angle valves (brass mounted), nipples, etc. The inlet enters on the side of the subsiding chamber somewhat above the bottom and is provided with a balanced float valve connected with a float on top of the filter to regulate the incoming water and keep the filter uniformly full of water. The waste valve is one of our special pattern, with square opening and flange cast to radius of main tank, to which it is secured by a wrought iron flange and bolts. The valve for cleaning the subsidence chamber is of the same pattern and put on in the same manner. The subsidence basin is also fitted with a large improved man-hole to enable one to enter, should it be desired to do so.

The pure water delivery is fitted with our "Automatic Controller" for maintaining a uniform rate of filtration. This device needs no attention whatever, and may be adjusted in a moment to any rate of delivery desired. An opening is provided in the cross on the outside of filter to which a steam connection can readily be made for the sterilization of the bed, whenever this becomes necessary. The filtering material is generally from three to four feet in depth and composed of specially mined white sand (almost pure silica). The wash and collecting strainers are made entirely of brass, with bronze screens and deflecting plates, and all are securely copper riveted. There are several hundred of these in each filter. The foundations required for these filters are most simple and durable, as no connections are taken from the bottom. All parts are easily accessible and are constructed with special care by skilled workmen; the material used is all first-class, and for durability and perfect operation, these filters cannot be excelled.

CONNECTING.

The natural or turbid water main or discharge from the low service pump for supplying the filters is connected to inlet valve 1. Valves 2, 3 and 6 are for washing or cleaning and are connected to a waste-pipe or led directly into open drain troughs or gutters. Valve 4 is connected with the pipe conveying filtered water for washing purposes. Valve 5 is directly connected to the "controller," which discharges the filtered water into the clear water basin, well, or reservoir, from which the main pumping engines take their supply. The steam connection for sterilization purposes is made to top of cross, to which is also attached the pressure gauge. Directly underneath this steam connection in the bottom of the cross is the sample or try cock.

When the filters are washed with the natural water, an extra connection from supply pipe to valve 4 is made. This is sometimes advantageous in saving filtered water, and can be used when the natural water is fairly good, or may be used for preliminary washing. In cases of fire or other causes which necessitate a sudden increase of capacity or output, the connection to rewash valve 3 may be easily shifted so as to also discharge filtered water into the clear water basin or reservoir. If it is desired to connect pure water valve 4 directly to a main discharge pipe, a Tee should be inserted in the main and an open branch or standpipe run up to the height of the filters so as to prevent an overdraft on them. Valves 2 and 6 may be placed at any point around the filter, and, in most instances, are placed so as to both dis-

charge into the same drain pocket or trough. Valve 1 may also be located at any desired point around the filter, convenient to the supply pipe. The girders or beams on top of the filters for carrying the stirring apparatus may also be placed at any angle desired, convenient to the line shafting, driving pulleys, etc.

WASHING.

To wash the filter, all valves having been previously closed, open valve 6 wide open. This allows all water above the inner tank and in the annular trough to drain out. Then open valve 4 slowly until sufficient water is on to allow the agitator to move easily. A trial pressure gauge may be put on the cross or wash pipe to show when sufficient pressure or volume of water is on for washing. This gauge should indicate from 6 to 10 pounds during washing. After a moment, or when the wash water has begun to flow freely over the trough, the stirring apparatus may be started on the forward motion. The rakes or bars then penetrate the semi-fluid bed and thoroughly agitate the same, sweeping all the sedimentary matter over the edge of the inner tank into the annular trough, from whence it is discharged through valve 6 to the sewer or back into the river. From 5 to 10 minutes is usually allowed and found to completely cleanse all of the filtering material. Washing is generally considered complete when the water resembles the natural or turbid water. It is not necessary or advantageous to continue the operation after the wash water has become white or slightly colored. When the bed is sufficiently clean, first reverse the motion of the stirring apparatus (while the current of wash water is still on) so that the rakes or bars will come out of the bed and rest on the surface, and while they are running idle or backward, as above, slowly close wash valve 4; then stop the agitator and close washout valve 6.

REWASHING.

After the filter has been washed, as above, open inlet valve 1 slowly, and when the filter is nearly full of water open rewash valve 3 about one or two turns; after a minute sample the effluent from trycock on bottom of cross, and when this becomes clear and satisfactory in appearance, close rewash valve 3 entirely; allow inlet valve 1 to remain full open. The object of "rewashing" or filtering to waste immediately after washing the filter is to displace any turbid or impure water remaining in the bed. It is obvious, therefore, that this operation is only imperative when the natural or turbid water is used for washing, and also that only a minute or two is required to completely displace the turbid waste with pure filtered water. When filter is washed with filtered water this operation is seldom found necessary. The circulation of the water in "rewashing" is almost identical with "filtering."

FILTERING.

After rewashing, simply open valve 5 slowly a turn or so to the required extent. If a "controller" is used, this valve may and should be opened full at once.

In this operation the turbid or impure water (having been previously charged with its modicum of the coagulant, if this is required) enters the subsidence basin through inlet valve 1, which is located a short distance above the bottom. The water then is slowly circulated around this basin by the deflecting flange on the inside directly in front of the inlet pipe. In this way the deposition of sediment is equal over the entire basin and the incoming water prevented from causing local or short-cut currents. Sedimentation takes place therefore, very rapidly in the almost quiet water, and before the water reaches the upper central discharge from the basin, all of the heavy matter and most of the finer impurities which have been coagulated are arrested. For illustration, samples of water that are very dark, brown or black in color, at the point of inlet, are almost white, or only slightly colored when finally discharged onto the filter bed. (It is variously estimated that from 50 to 75 per cent. by weight of the suspended impurities are caught in this basin alone.) Tests have actually proven that from 30 to 70 per cent. of bacteria are arrested or retained in these settling basins. The great saving in capacity, space,

33

and first cost, as well as in the maintenance of a Jewell filtration plant is, therefore, quite obvious. It is understood that in this process the coarser impurities are retained in the basin without coagulation. Another advantage worthy of note here is that during this long opportunity afforded for subsidence, **the coagulant, if any is used, becomes completely precipitated, not a trace being found in solution when the water reaches the filter bed.** It is also apparent that this thorough reaction likewise effects a great saving in the reagent over the old method of introducing it as the water is being discharged directly on the bed; and with this improvement there is absolutely no possibility of any coagulant getting through with the filtered water. The water, after subsidence, leaves this basin and rises through the central standpipe in the filter bed, overflowing it. A low head of water (about two feet) is, however, carried on the bed so as to prevent the incoming water cutting channels or furrows in the bed. The current of the water then proceeds downward through the little interstices of the bed, depositing thereon whatever impurities have resisted sedimentation, and thence is collected evenly and proportionately from all parts of the bed by the strainer system, first entering the numerous strainers themselves, then the parallel branch pipes and manifolds and, finally, coming together in one volume in the

"downdraft" pipe, which runs directly to the outside of the filter communicating with the cross to which valves 3, 4 and 5 are attached, from which latter valve it is discharged clear and bright.

SURFACE AGITATION.

It is a well-known fact that the greater percentage of the impurities are retained upon and in the first half-inch of the beds of filtering material; and this percentage obviously increases with the amount of coagulant used and heavy sedimentary matter carried by the water. Numerous attempts have been made to increase the economy of various processes by removing this deposit without impairing the other portions of the bed, or quality of the effluent, or necessitate the washing of the entire bed, but all have signally failed.

With the introduction of our new gravity filter, however, this problem was solved in a most practical and efficacious manner, and the saving in wash water alone has been estimated at from 40 to 60 per cent. This result is accomplished by plowing or raking the surface of the bed by means of the reversible agitator. The surface accumulation or film, as it is called, is thereby broken up into irregular shaped lumps or masses which are allowed to remain, and the incoming water then freely passes around and between them to the comparatively clean sub-surface.

The operation requires but two or three minutes on the largest filters and may be advantageously repeated two or three times between washings of the filter. Filtration is momentarily stopped by simply closing the outlet valve 5; then the agitator is revolved slowly on the reverse motion from one-half to one turn. **The pure water valve** 5 may then be immediately opened, slowly, and not even the slightest trace of sediment will appear, the water being in every particular equal to that filtered through the sediment layer.

The principle underlying this peculiar action of our new filter has not as yet been fully explained, but undoubtedly lies for the most part in the action of the "downdraft" pipe on the filter bed. In filters operating under a head or pressure of water above the filtering beds, surface agitation is wholly impracticable, due to the sediment being carried through with the filtered water.

CLEANING THE SUBSIDENCE BASIN.

Probably no one feature of the improved filter is more ingenious or novel than the method of washing out the heavy sediment or sludge which so rapidly accumulates in this basin, for which reason every possible facility is included and applied for hastening the operation, lessening the labor and insuring a thorough cleansing. First of all, this operation is carried on simultaneously with the washing of the filter bed; and, second,

the same wash water thoroughly cleanses both the filter bed and subsidence basin; third, no extra time or labor is required. The operation is automatic, and as for efficiency and cleanliness, nothing more could be desired. In cleansing the basin, valve 2 (instead of valve 6) is opened full and then the filter washed in the usual manner above described. As valve 6 has remained closed, the dirty wash water, instead of overflowing the inner tank and escaping via valve 6, is compelled to rise a few inches higher and then overflows down through the central standpipe into the subsidence basin. At the bottom of this standpipe, the falling water comes in contact with a large open deflecting elbow, which is fastened to the same vertical shaft which carries the agitator or rakes, and, therefore, revolves with it, and the water is, therefore, thrown with considerable violence against the lower sides of the basin, thence sweeping rapidly backward in almost a torrent towards the center, carrying along with it even the heaviest accumulations of quick-sand or silt, from which all are discharged out to waste through valve 2. Accumulations of **solid sediment over a foot deep** are washed out with the greatest ease and economy; no extra water being required; and, on removing the man-hole at the side, we have invariably found the floor so clean that one could scarcely soil his new linen on it.

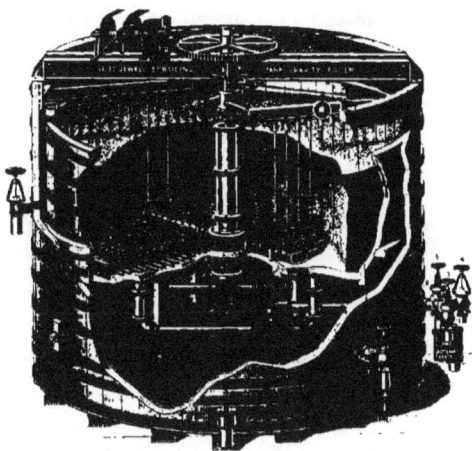

Seventeen Foot Jewell Subsidence Gravity Filter.

Description of Seventeen Foot Jewell Subsidence Gravity Filter.

UR new seventeen-foot Gravity Filter, il- lustrated on opposite page, embodies the same general principles of construction as have proven so successful in the smaller sizes.

The wash and collecting screen system in this filter is perfectly proportioned and every part of this large bed is lifted and floated uniformly. This equal distribution is afforded by four large branch manifold lines, each radiating from the central section, into which are screwed the extra heavy distributing pipes that carry the equalizing screens, and each of the manifold branches re- ceives a separate branch pipe which later con- verge into one large outlet.

The stirring apparatus deserves especial men- tion from its many points of excellence and in- genuity. The weight of the apparatus is mainly carried on the upper edge of the inner tank, the balance being carried upon two heavy "I" beams extending across the main tank. The driving mechanism is par excellence in every detail. The central vertical shaft, which carries the rakes or

stirring bars is keyed to a large and heavy spur gear, which engages a small pinion that is fixed to a large bevel gear, and the latter is driven by a small bevel pinion mounted on the end of the horizontal shaft and which also carries two fric- tion clutch driving pulleys. A heavy cast iron saddle frame, of one piece throughout, is secured to the "I" beams adjacent to the main gear wheel and holds both bevel gears in perfect running con- tact. All boxes are either heavily babbitted or lined with brass bushings. The friction clutch pulleys are arranged in duplex and operated with one shifting lever, which extends to and rests upon the edge of the main tank. The pulley for driv- ing during washing is six inches in width, being twice the size of the reversing pulley. The upper edge of the inner tank is covered with sectional railing to form a track upon which ride the whee's carrying the "I" beam cross arms of the agitator.

Aside from the modifications above mentioned there are no other practical differences in con- struction from the smaller sizes, and the same methods of operation apply to all sizes.

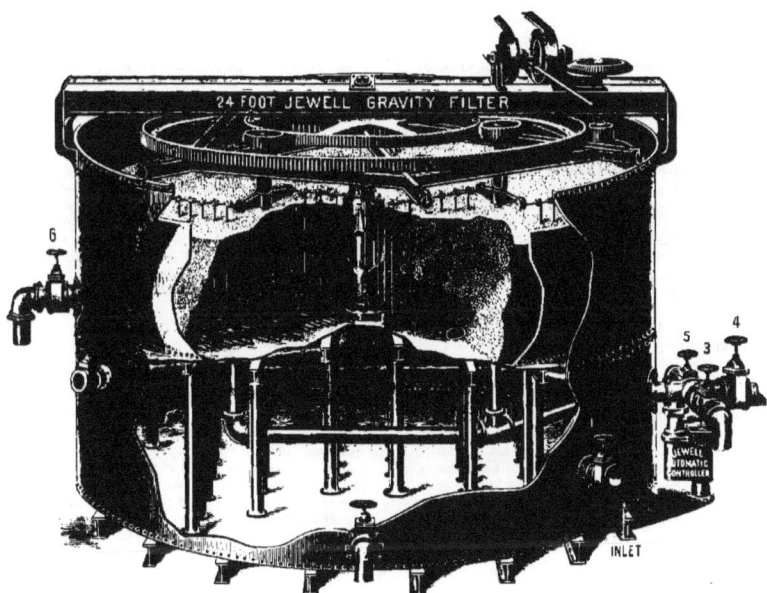

Twenty-Four Foot Jewell Subsidence Gravity Filter.

Description of Twenty-Four Foot Filter.

THE 24-foot Jewell Gravity Filter, as shown on page 38, is no doubt the largest mechanical filter ever built in the world. It has a filtering area of 452 square feet, and a settling basin situated below the filter-bed, which contains 23,000 gallons of water; and will have a capacity of easily purifying an average of 1,000,000 gallons per day of 24 hours, of river waters in their ordinary stages. The filter-bed holding tank contains 1,808 cubic feet of filtering material, and the bed is 4 feet in depth.

This machine combines many novel features never before applied to mechanical filters. It operates on the down draft principle, and is provided with controllers for both the inlet and purified water, thereby insuring a constant and uniform delivery. The construction is such that the filter-bed holding tank can be washed independently of the settling basin, or simultaneously, as desired. This agitator is operated by gearing as shown in cut; the center gear being a fixed and stationary arrangement. The outer gear is securely fastened to a rack having six projecting arms which rest with rollers on the upper rim of the outside tank, as shown in cut; this gear being driven by spur and bevel gears. The three gears carrying the agitator arms are revolved by moving around the central fixed gear. This permits of the agitators being revolved at the same time they are traveling around the entire circumference of the tank in a circular motion, thereby reaching every portion of the filter-bed. The bars of the agitator which project into the filter-bed are hinged, so that when the bed is being washed they stand in a perpendicular position. After washing, the agitator is reversed, and these bars immediately assume a horizontal position and lie on the surface of the bed, and can be drawn over this surface when desired, which will serve to break up the film or crust which may have formed on the surface of the bed, thereby permitting the filter to assume almost the same conditions as after washing. This surface agitation may be carried on from two to four times before it will be necessary to wash the filter-bed; thus effecting a great saving in time and in the amount of water used for cleansing the bed.

The tanks in this filter are preferably built of steel, and the filter-bed holding tank supported on cast iron columns and I beams, as shown in cut. This filter is designed more especially for large cities and where a large amount of water is to be purified. The fact of the large filtering area very much lessens the ground area and cost.

TABLE OF STANDARD SIZES
OF THE
JEWELL SUBSIDENCE GRAVITY FILTER.
With Separate Settling and Filtering Compartments.

No.	Size, Diameter in Feet		Connections in Inches		Capacity. Minimum and Maximum, U. S. Gallons.				Area.	Bed.	Shipping Weights (Approximate.)			Total Weight
For Convenience in Telegraphing	Filter Bed Tank. (Inside.)	Main Tank. (Outside.)	Supply and Discharge Pipe.	Washout Pipes.	Minute.	Hour.	Day: 24 Hours.	Settling Basin.	Effective Filtering Surface.	Filtering Material.	Machine Work.	Tank Material Cypress or Cedar.	Filtering Material.	Filter in Operation.
									SQ. FT.	CU. FT.	LBS.	CWT.	TONS	About TONS.
1	6	7	3	6	47—94	2,800 - 5,600	62,500— 125,000	1,500	24	113	1,000	50	5	15
2	8	9¼	4	6	82— 164	5,000 - 10,000	120,000— 240,000	2,800	50	200	1,800	70	9	24
3	10	11½	4	8	130 - 260	7,800—15,600	185,000— 370,000	4,000	78	312	2,500	105	14	45
4	12	13½	5	8	188 - 376	11,300 - 22,600	250,000 - 500,000	5,750	113	452	3,750	125	20	70
5	14	15½	6	8	255— 510	15,300 30,600	365,000 - 730,000	8,000	153	612	5,000	145	27	100
6	17	18½	8	8	376 - 732	22,800—45,200	500,000—1,000,000	11,500	231	904	9,500	180	40	145
7	21	22½	8	2-8	565—1,130	33,900—67,800	750,000—1,500,000	17,000	339	1,356	12,000	230	80	200
8	24	26	10	2-10	753—1,506	45,200—90,400	1,000,000 2,000,000	23,000	452	1,808	17,000	300	80	295

Standard Height of Filters, 14 feet.   Depth of Filter Beds, 4 feet.

Special Sizes and Modifications made at Regular Prices.   Tanks of Steel, Iron or Masonry, if preferred.

Jewell Gravity Filter Plant, Erected for the Wilkes-Barre Water Company.
Capacity, 10,000,000 Gallons daily.   Exterior view of Filter House, Engine and Boiler Rooms.

WILKES-BARRE, PA.

Jewell Gravity Filter Plant, Erected for the Wilkes-Barre Water Company.
Capacity, 10,000,000 Gallons daily.   View of Filter House and Clear Water Reservoir.

42

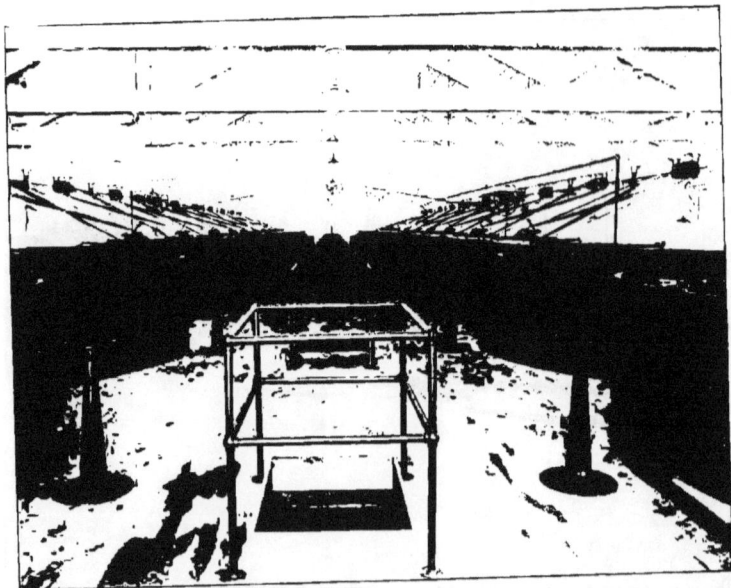

Jewell Gravity Filter Plant, Erected for the Wilkes-Barre Water Company.
Capacity, 10,000,000 Gallons daily.   Upper Section of Filter House and Filters.

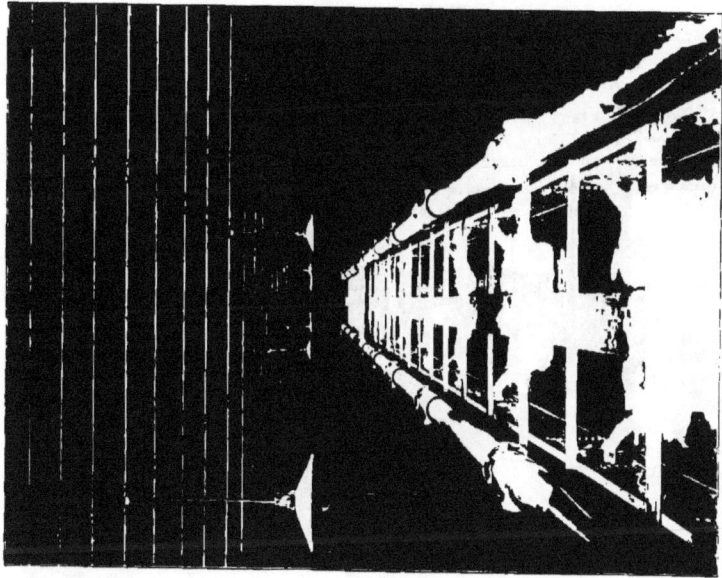

Jewell Gravity Filter Plant, Erected for the Wilkes-Barre Water Company.
View Showing Clear Water Flume and Wash Water Mains, looking
between the two Rows of Filter Tanks.

Jowell Gravity Filter Plant, Erected for the Wilkes-Barre Water Company.
Capacity, 10,000,000 Gallons daily.   Engine and Dynamo Room.

Jewell Gravity Filter Plant. Erected for the Wilkes-Barre Water Company.
Capacity, 10,000,000 Gallons daily.
Apparatus for the preparation and use of Alumina Solutions.

The Jewell Gravity Filters at the Water Works of Wilkes-Barre, Pa.

THE works of the Wilkes-Barre Water Company were built in 1857-9. The company draws a gravity supply from impounding reservoirs, the combined storage capacity of which is over 2,000,000,000 gallons. Prior to the completion of a large new reservoir the gravity supply was supplemented by pumping from the Susquehanna River. The new reservoir has a storage capacity of about 2,000,-000,000 gallons, and was put in use in August, 1891. The reservoir covers nearly 400 acres and has an average depth of 15 1-3 feet. Its drainage area is about three miles above the filter plant, to reach which the water flows down the channel of a natural stream, falling nearly 300 feet.

The first summer the reservoir was in use trouble was caused by algae growths, beginning in July and reaching its height in September, the algae forming a green scum on the surface of the reservoir and even showing in the city. The water was described as having a fishy odor and taste.

THE 10,000,000-GALLON JEWELL FILTER PLANT.

After operating an experimental plant the Wilkes-Barre Water Company contracted with the Morison-Jewell Filtration Company for gravity mechanical filters with a guaranteed daily capacity of 10,000,000 gallons. The plant was tested the latter part of July, being run at the rate of 11,100,000 gallons per day, and was put in operation Aug. 1, 1895. Mr. Edmund B. Weston, M. Am. Soc. C. E., of Providence, R. I., was Consulting Engineer to the company and prepared all the general plans for the plant.

The water comes by gravity to the filter house from either one or both of two small pools a few feet distant, these now serving as settling basins, although having been built for other purposes. The water is screened as it leaves the pools and flows to the filter house through a 30-inch pipe.

The filter house is of brick, the main building being 51 2-3x181 2-3 feet, over all, and the engine and boiler room 34x51 2-3 feet, over all. The filter tanks are supported on brick foundations, beneath which is a 2-foot concrete floor which extends over the whole of the basement. All inside brick walls and the roof trusses are whitewashed; the roof ceiling, clear water flume and feed pipe for wash water are painted white.

The unfiltered water comes first to the settling tanks, two on each side, where it receives the basic sulphate of alumina used as a coagulant. It then passes through a 24-inch supply main on the outer side of each row of tanks, with branch connections to the several filters, the water entering above the filtering material and being spread by the deflectors. The deflectors are designed to aerate the water and at the same time prevent the plowing up or packing of the sand by the water when it strikes the beds. They are attached to the vertical agitator shaft.

The filter tanks are made of cypress, the staves here being 18 feet long, an extra 4 feet having been added to the height of the tank on account of local conditions. Each tank has a filtering area of 113 square feet, giving 2,260 square feet for the twenty filter tanks, four valves are used to control the flow of water to and from the tanks. These are all operated from the upper floor and are placed side by side and labeled in the following order from left to right: "Sewer," "Rewash," "Wash," "Filtered." Each filter tank is numbered and has an electric push button by means of which, and the proper connections, signals are transmitted to the engine room for starting and stopping the machinery used in washing.

The maximum amount of alum used is given as ¾ grains per gallon. At times none is used, it being unnecessary, for instance, when the ground is frozen. The alum is mixed in two tanks 8 feet in diameter and 10 feet high. One tank is drawn upon while the alum is being mixed in another. The alum is placed on a rack at the top of the tank, and as the water takes up the alum the greater specific gravity of the latter causes it to fall to the bottom. To keep the solution all of the same specific gravity after mixing, compressed air is used to stir it up. The alum is pumped to the settling tanks by phosphor bronze pumps. The gate on the supply main keeps the influent at a fixed volume and the amount of alum is varied by altering the speed of the pump.

Washing is done with filtered water and once in 24 hours, ordinarily, no matter how clear the water may be. The wash water is turned on to loosen the sand before starting the agitator, or revolving rake. The agitator revolves 8 to 10 times per minute, and was run for six minutes on a tank which the writer saw washed. The wash water could be seen bubbling up through the sand bed when it was first turned on. The turbidity of the water in times of flood is in part due to the fact that before reaching the filters the water flows about three miles in the bed of a stream with 300 ft. fall. High hills rise at each side of the stream and a country road passes along side of it, the wash into the stream being great.

The equipment of the boiler and engine room is as follows: There are two 50 H. P. boilers, made by the Vulcan Iron Works, of Wilkes-Barre. The boilers are equipped with automatic damper regulators. Both boilers were in use at the time of the writer's visit in March, but it was hoped that one would be sufficient when steam for heating was no longer necessary. A 10x10x12-inch Blake pump, with a capacity of 4.08 gallons per stroke, is used to lift filtered water for washing. The pump does fair work at 110 to 120 strokes per minute, and the attendants are instructed to keep it up to 100 strokes. On the average the pump is operated at 110 strokes per minute, for from 7 to 10 minutes, to wash one filter, giving from 3,100 to 4,500 gallons per wash per tank. The water for washing is raised about 13 feet. A 15 HP. Taylor engine is used to drive the sand agitators, and there is a 6x10x12-inch Blake air compressor for agitating the solution in the coagulant tank.

The building is lighted by 64 incandescent 24-candle power lamps, and the grounds by four arc lamps installed by Hessel & Lewis, of Wilkes Barre. The dynamo was made by the General Electric Co., and is driven by a Payne automatic engine.

On March 2, 1896, the force employed at the filter plant was reported by Mr. McGarry as one engineer, one fireman and one washer, in two shifts of twelve hours, making six men in all. One of the men rated as fireman regulated the gates at the impounding reservoir, three miles distant, and changed the screens at the gate house of the pools at the filter house. On one day in February, 1896, the screens had to be changed once an hour for 15 hours on account of the wash from the adjacent country. On the same date Mr. McGarry informed the writer that the cost of filtering 7,000,000 gallons in 24 hours, on Feb. 21, 1896, was $16.15, divided as follows:

Two engineers at $2.15	$4.30
Two firemen at $1.75	3.50
Two laborers at $1.50	3.00
Coal	.78
Hauling coal	.75
250 lbs. alum at 1¾ cents	3.82
Total	$16.15

The amount filtered could be increased to 10,-000,000 gallons without any additional labor.

Below is a letter, dated April 29, 1896, from Mr. McGarry, to Mr. John C. Trautwine, Assoc. Am. Soc. C. E., Chief of the Philadelphia Bureau of Water, in which figures are given for the cost of filtration under different circumstances, and some other information regarding the plant is furnished:

"In reply to your inquiry regarding cost of filter plant will say that the total cost of constructing the same has amounted to $122,361.79.

"At present we are filtering a little over 9,000,000 gallons daily, at an expense as follows:

Two engineers, at $2.15 per diem	$4.30
Two washers, at $1.62½ per diem	3.25
Fuel	1.30
Oil waste, etc.	.11
Total	$8.96

"Regarding the results of filtration, will state that it is very satisfactory. The use of coagulant is only intermittent with us, using it only when the water is very turbid, and when it holds in suspension heavy deposits of loam which is carried into the creek. We have not used any chemical since April 7—nothing but quartz filtration, and the quality of the water is highly improved. A slight stain remains on our supply (without the use of any solution) which an expert might detect. I may add that we have had a run of 91 days, and during this continuous run we used on an average half a grain sulphate of alumina per gallon."

Jewell Gravity Filter Plant, Erected for the Cataract Construction Company.
Capacity, 4,500,000 Gallons daily.   Purifying Niagara River Water.
General view of Filter Buildings.

51

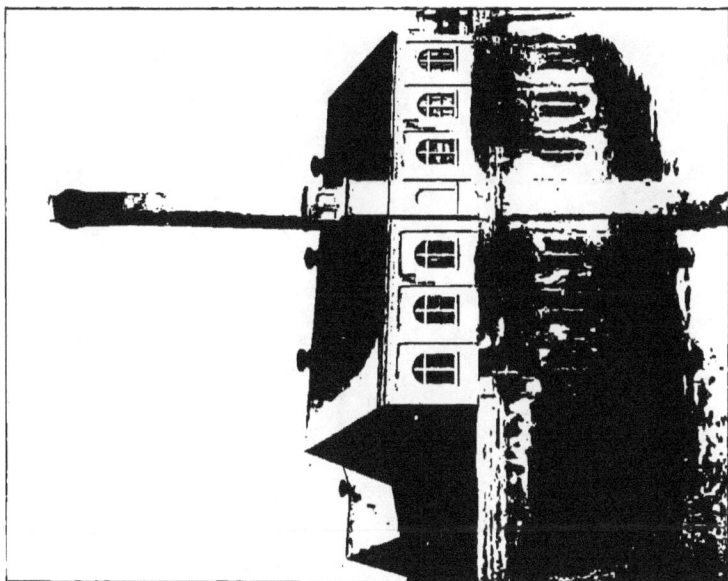

Jewell Gravity Filter Plant, Erected for the Cataract Construction Company.
Capacity, 4,500,000 Gallons daily.   Purifying Niagara River Water.
View of Filter House and Canal.

Jewell Gravity Filter Plant, at Niagara Falls, N. Y.   View from Upper Section.

Filter Plant, Erected for the Cataract Construction Company. 4,500,000 Gallons daily. Purifying Niagara River Water. Entrance to Upper Section of Filter Room.

The Jewell Filter Plant at Niagara Falls, N. Y.

THE population of Niagara Falls in 1890 was 5,502. The population supplied by the Company in May, 1896, was estimated as 9,000. The water supply was introduced in 1877, and the works are now owned by the Niagara Falls Water-Works Company, which is controlled by the Cataract Construction Company. Water is taken from the Niagara River, the pumping station and the filter house being located on the west side of the canal of the Niagara Falls Power Co., as shown in view on page 52. Another exterior view showing the front is presented on page 51. Gravity filters are used, the average extra lift being 14 feet. The working head averages only 4 feet.

Filtration was adopted "for the removal of suspended and organic impurities." In washing, filtered water under pressure from the mains is used; the filters being washed on a average of about once in twenty-four hours. The clear water reservoir has a capacity of 30,500 gallons, in addition to which there is a 750,000-gallon standpipe. Interior views of the plant are shown on pages 53 and 54.

The filtration plant is contained in an annex adjoining the old pumping plant, but new buildings were constructed in 1895 of pressed grey brick with ornamental stone facings to cover both the pumping station and the filter house. The supply, as above stated, is taken from a canal leading from the Niagara River, the same canal also being the supply for the electric power development of the Cataract Construction Company. The incoming water is admitted through screens to a well under the pumping station, and from there is conducted through a 24-inch pipe to two large settling basins so constructed with partitions as to have the movement of water in them as slow as possible. These settling basins are also united by equalizing pipes so that an even level may be preserved in both. From the last section of the settling basins the water is lifted to the filters by two centrifugal pumps, electrically operated, each pump having a daily capacity of 5,000,000 gallons. The filter plant proper consists of nine (9) Jewell Gravity Filters, each 13 ft. 8 in. outside diameter by 15 ft. 6 in. high. The water is admitted to the filters from the centrifugal pumps over Thurston aerators. The filter beds are composed of white machine crushed quartz 3 feet in thickness. The filtered water first passes to a clear water reservoir below the filters, and from this reservoir the water is then taken by the main pumping engines and delivered to a stand-pipe. The pumping plant consists of two Snow High-duty Duplex Pumps, 17-inch cylinders with 18-inch stroke, and are located on by-passes so as to be able to pump either filtered water or river water.

Since the plant was started in operation, April 1st, 1896, none but filtered water has been supplied to the city. The average daily pumping capacity is now 3,000,000 gallons. Basic sulphate of alumina is used as a coagulant, the average quantity used being ½ grain per gallon.

LORAIN, OHIO.

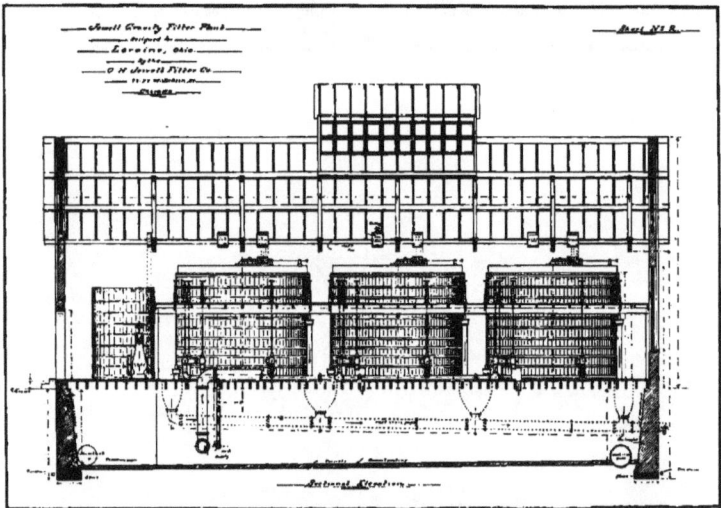

Plant of Seventeen-Foot Jewell Gravity Filters at the Water-Works Pumping Station.
Erected for the City of Lorain, Ohio. Capacity, 3,000,000 Gallons daily. Purifying Lake Erie Water.
Elevation view of Filters and Section of Building.

LORAIN, OHIO.

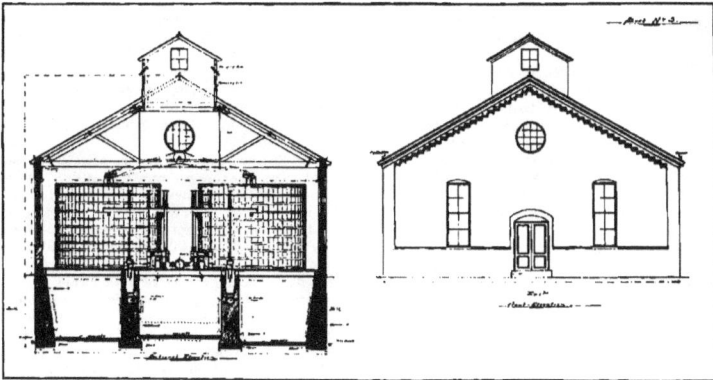

Plant of Seventeen-Foot Jewell Gravity Filters at the Water-Works Pumping Station.
Erected for the City of Lorain, Ohio. Capacity, 3,000,000 Gallons daily. Purifying Lake Erie Water.
End view of Filters and Building.

Description of the Jewell Gravity Filter Plant at Lorain, O.

THE contract for this plant was only entered into in May, 1896, and at the present writing the plant is under construction. The tanks of these filters are larger than any heretofore built, having a filtering area 17 feet in diameter, while the outside diameter of the main tank is 18⅔ feet. A sectional cut of this filter is shown on page 36, and a detail description of same is given on page 37.

The filters are arranged in two rows of three each, directly over a clear water basin. This basin is divided into three parts by two heavy walls, running lengthwise of the building, upon which the filters are mainly supported. These walls are provided with drain pockets and pipes for carrying off the wash water, as shown on page 58, and in dotted lines on page 57. Conduits are provided in the two dividing walls of the reservoir, so that a constant circulation of water is maintained in all parts of same, and the floors of each compartment are graded to facilitate this.

The main feeder for the filters lies upon the floor of the filter house, as does also the wash water pipes; in fact, there are no pipes below the floor—a feature which will be readily appreciated. The valves are all operated from a platform or gallery, running almost the entire length of the building. This platform is situated about 4 feet below the top of the filters, thereby affording a convenient view, and greatly facilitating the operation of the filters.

The line shaft for driving the agitators is carried upon the lower horizontal girders of the truss roof. The engine, which is located on the floor of the filter house, is of the vertical type, and drives directly onto the line shaft. The tanks for the use of alumina sulphate, of which there are two, are located on the same level with the filters. They are connected in duplicate so as to work alternately. The alumina solution is drawn from these tanks by a small steam pump, which is located in the pump house and arranged to work automatically by the main supply pumps. The solution is then discharged into the supply main leading to the filters.

The works at Lorain are owned by the city, and were built in 1884.

JEWELL
GRAVITY FILTERS

CAPACITY,
3,500,000 GALLONS
DAILY.

At the Pumping Station
of the
Burlington Water Company,
Burlington, Iowa.

PURIFYING
MISSISSIPPI RIVER
WATER.

JEWELL
GRAVITY FILTERS
at the Pumping
Station of the
Burlington Water Co.,
Burlington, Iowa.

CAPACITY,
3,500,000 GALLONS
DAILY.

Purifying Water
from the
MISSISSIPPI RIVER.

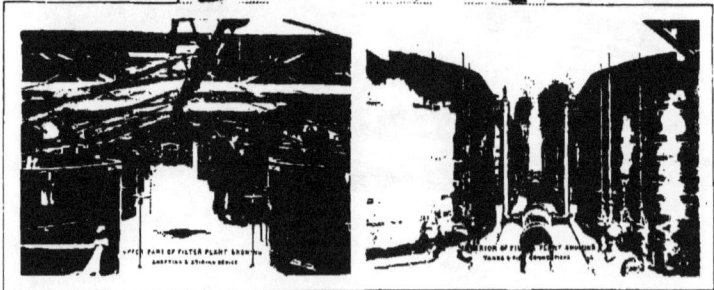

INTERIOR VIEW FILTER PLANT SHOWING
STAIRCASE TO PLATFORM

UPPER PART OF FILTER PLANT SHOWING
SHAFTING & STIRRING DEVICE

INTERIOR OF FILTER PLANT SHOWING
TANKS & PIPE CONNECTIONS

Jewell Filter Plant at Burlington, Ia.

[Extracts from *Burlington* (Iowa) *Hawkeye*, April 8, 1891.]

THE Water Company set about the erection of a filter plant last fall, and before winter set in had erected a filter house 40x184 feet and was ready then to continue, under cover, the installation of the plant. Previous to this, however, and even long before the court gave a judicial stamp to the obligation of the company to filter the water, President Rand and Superintendent Charles Hood had anticipated such an addition to the power plant and had prepared themselves by personal inspection of many various filter processes, and by careful study of the conditions of the service in Burlington, so that when the time came to act they had pretty well made up their minds what would best answer the purpose.

THE GRAVITY SYSTEM.

They decided that the gravity system of filtration as exemplified in the O. H. Jewell Filter Co., of Chicago, would best meet all requirements of the somewhat complicated problem. This, in brief, consists in the pumping of water from the river into immense cypress tanks containing a bed of ground quartz, through which the water passes cleared of all impurities, into a clear water basin, from which it is pumped through the mains. This, it will be seen, requires two separate and distinct pumping operations. The new 6,000,000 gallon power pump is continued on the mains, drawing from the clear water basin instead of the river direct, as heretofore, and the original Holly quadruplex pump was converted by the addition of four water cylinders from a high-service pump of 3,000,000 gallons to a low-service pump of 12,000,000 gallons, and this was given the duty of pumping from the river into the tanks. Illustration No. 3 on page 61 shows the interior of the filter house from the south end. The rows of tanks are shown on either hand, and the big pipe that passes between is the main feeder from which branches lead into each tank. There are now eleven of these big tanks that are made of the best quality of Louisiana cypress. The tanks are 13 feet in diameter and 14 feet high. They rest upon heavy masonry running lengthwise through the clear water basin and dividing it into three compartments connected by sluiceways. This latter has a capacity of 500,000 gallons, and the eleven tanks now in operation are capable of filtering, under ordinary pressure, 3,500,000 gallons daily. Provision is made for future needs of the service in the space allowed for the erection of nine additional tanks by which the capacity of the plant will be nearly doubled. Illustration No. 2, on page 61, also an interior view from the north end of the building, shows the space in reserve and also makes clear the relation between the tanks and the clear water basin. The square apertures in the dividing walls are found beneath all the tanks; at a depth of about 4 feet from the top a big pipe runs from end to end and leads to the sewer. The purpose of the apertures will be explained later. Within the outer tank is one of smaller diameter leaving a space of about 4 inches all around it. In the smaller tank is the filter-bed proper. It first consists of 400 aluminum bronze screens, each screen having a deflecting plate and all substantially riveted. On top of the screens is the 2½-foot of machine crushed quartz of the fineness of very coarse sand. Each filter has a car load of this material, and each is filled with Jewell's latest improved agitators for stirring the quartz bed during the operation of washing out impurities collected in the process of filtration. A line of shafting runs the full length of the building on either side, and to it the agitators are geared so that any one may be used independently of all others.

Now, a woman would not think of washing a piece of dainty lace in dirty water—no more would Superintendent Hood think of washing his filters with dirty water. When he wants to clean them he shuts the valve that admits the water from the river and starts a new low-service pump in the southwest corner of the power-house. This has a capacity of 1,500,000 gallons and draws from the clear water basin. Instead of flowing into the top of the filter tank, however, the water is forced up from the bottom, through the screens and through the quartz bed. The agitator is now set in motion and the long rake-like arrangement goes round and round in the quartz, thoroughly stirring it and allowing all impurities to rise with the flow of water from beneath. This flows over the rim of the inner tank and into the square apertures before referred to in the dividing wall of the clear water basin, whence it is carried off into the sewer through the pipe cemented in the wall. The arrangement for carrying off the wash water is very neat in design and the large cumbrous pipes usually adopted for this purpose are avoided.

The platform shown in the interior view is erected about 12 feet above the lower floor line, and all the valves are operated therefrom by means of rods extending upward through the floor, so that the entire plant is easily controlled by one man, who has a perfect oversight of the working of each filter, working singly or in battery.

The water which has gone through the process of filtration above described will be practically pure, and absolutely colorless at all times, as now, when it has a faintly yellowish tinge or stain, which is ascribed to the influx of the streams and feeders in the pineries in which the pine cones and needles have been steeped. Even this stain can, however, be removed by a part of the plant which has not yet been described, but which will now engage our attention for a few moments.

COAGULANT PLANT.

The coagulant plant is contained in a small brick annex to the pump house and another new pump is required for its operation. But this, the fourth pump in the plant, is a very small concern—it occupies merely the space of a window sill. It draws from the alum tanks and jets into the main artery leading into the settling compartment of the filter tanks a solution of alum, the strength of which can be as easily and accurately determined as if required for compounding a physician's prescription. There are two tanks in which the solution is made from a very hard and crude form of alum. It is so hard that it is indissoluble in still water and requires the erosive power of running water to produce any effect upon it. These tanks are connected by a "Y" with the little pump and, valves being placed at certain points, the solution may be drawn from one or the other as desired. It is not the desire of the company to make use of the coagulant at all times, and during the summer it is thought it will not be required at all. It has been taken for granted that the reader understands that a very small percentage of alum will immediately clarify a large quantity of water by pre-

cipitating all impurities contained in it. With this in mind the work of the little pump will be understood.

It will interest Hawk-Eye readers to know that, except in the event of a general conflagration, nothing but filtered water will hereafter be sent through the mains. The terms of the charter do not require this in case of even ordinary fires, but the capacity of the plant is ample for anything short of a holocaust. With 500,000 gallons to start on a supply can be maintained for a long-continued and heavy demand and, should it become necessary, the capacity of the filters can be increased by making the filtration less thorough, and consequently more rapid. Hereafter we will not only drink, bathe and cook with filtered water, but we will sprinkle our streets and lawns and extinguish our fires with clear, odorless, colorless and sparkling water.

To Superintendent Charles Hood, of the Water Company, belongs the credit for the installation of the plant. He has made a thorough study of the filter question and is now one of the best authorities in the country on the subject. During all the years the subject was debated and wrangled over he was quietly gathering together a mass of facts and the experience of others, and digesting the material thus obtained. It was all along his personal desire to see a filter plant installed, and when the time came to do so he was ready for the work and took a keen personal interest in the realization of his hopes and plans. He takes a pride in knowing that he has now under his charge as complete and satisfactory a system of water works as there is in the country.

The Jewell people have taken unusual interest in the Burlington plant, expecting, as they do, to refer to it as a model of their system. Mr. Jewell has given the best portion of his life to the study of scientific and practical filtration of water. Personally he is a man who inspires confidence, and the Water Company feels that his company has done its work well, complying in every respect with the very letter of the contract.

COST OF PLANT.

About $75,000 has been expended by the Water Company in the construction of the filter plant. Of this $33,000 represents the contract price to the O. H. Jewell Filter Company, and the balance is to be apportioned among cost of construction of buildings, superstructure and clear water basin, low-service pumps, chemical plant, etc.

LEXINGTON, KY.

PUMPING STATION.   FILTER HOUSE.   CLEAR WATER RESERVOIR.

Jewell Gravity Filter Plant, Erected for the Lexington Hydraulic and Manufacturing Company.
Capacity, 2,000,000 Gallons daily.   Purifying Surface Drainage Water.

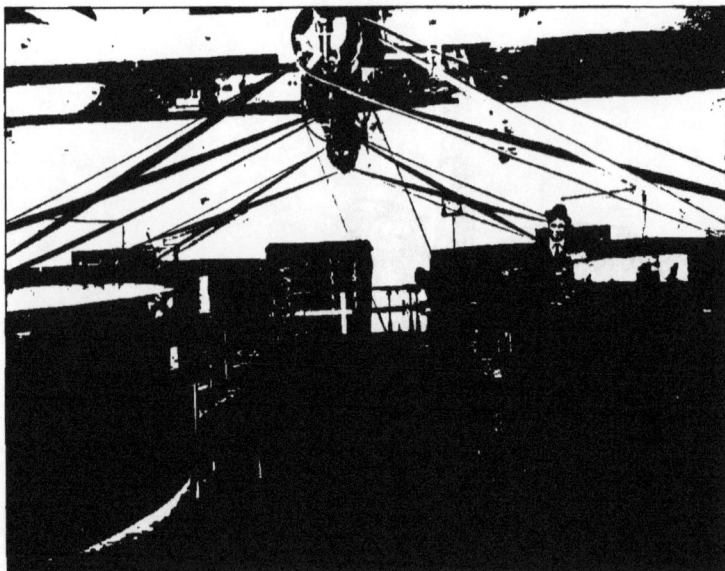

Jewell Gravity Filter Plant, erected for the Lexington Hydraulic and Manufacturing Company.
Capacity, 2,000,000 Gallons daily.   Purifying Surface Drainage Water.   View above Platform.

LEXINGTON, KY.

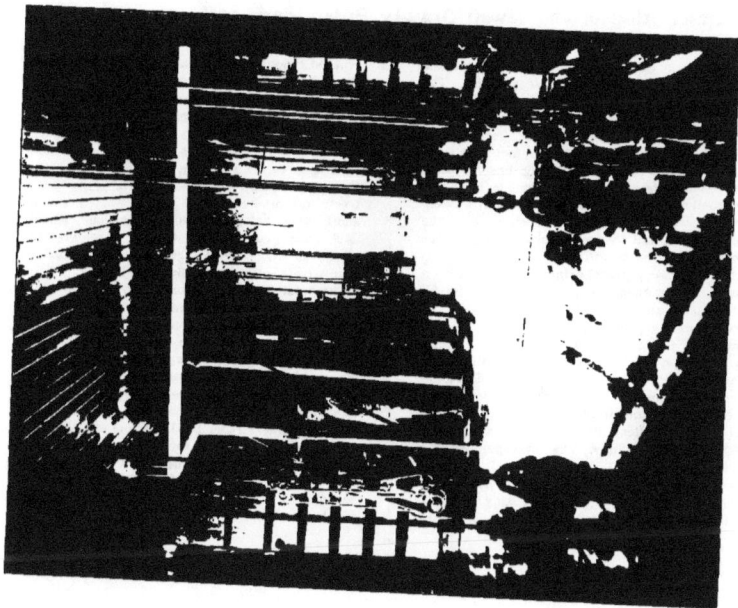

Jewell Gravity Filter Plant, erected for the Lexington Hydraulic and
Manufacturing Company. Purifying Surface Drainage Water.
Capacity, 2,000,000 Gallons daily. View below Platform.

Description of the Jewell Gravity Filter Plant at Lexington, Ky.

[From *Engineering News*, May 28, 1896.]

NE of the most interesting filter plants in the country is at Lexington, Ky. The population of Lexington was 21,567 in 1890. It has had a water supply since 1832, the present plant being owned by the Lexington Hydraulic and Manufacturing Company, of which Mr. S. A. Charles has for some time been secretary and superintendent, and is now receiver. Mr. Charles has also served as engineer for the company in connection with the filtering plant, and we are indebted to him for the information given below.

The water supply of Lexington is now drawn from a 360,000,000-gallon reservoir and, after filtration, it is pumped directly to the mains. The gravity filtering plant is located in a building adjoining the pump house. Filtration was adopted "to remove algae and vegetable growths, principally, also occasionally some mud," and necessitates extra pumping to a height of 20 feet. The average daily amount of water filtered in 1895 was 1,230,000 gallons, the maximum having been 1,550,000 gallons a day. The average working head is 4 feet.

The filtering material is white sand from New Orleans. From March 1, 1895, to May 15, 1896, the loss of sand has been 4 inches. The filters are washed twice or more a day, both filtered and unfiltered water being used, and 2 per cent. of the total quantity of water (unfiltered) delivered by the low-service pump, being used in washing. None of the wash water is lost, all going through a settling basin and coarse filtering process and then back to the reservoir. The capacity of the low-service pump is ample, so water is used freely for washing, some being wasted for a short time before resuming filtration. About double the quantity of wash water is required to clean the filters when algae are giving trouble than when the chief difficulty is mud.

The total amount of alum used in 1895 was 16,000 pounds, ranging from 0.6 grains per gallon for a short period, to 0.3 grain for a long time, and averaging 0.4 grains. The alum is bought from Harrison Bros. & Co., Chicago, at a cost of 2 cents per pound, delivered. No other chemical has been used as a coagulant. (Lime is now, however, being used in conjunction with the alum, giving better and more economical results.)

The complete cost of the filtration plant and accessories was $27,000. The only addition to the operating expenses caused by filtration is on account of the extra pumping, labor and alum, which was $2,193 in 1895.

The filtered water passes to a 333,000-gallon clear water reservoir, covered with a deck roof and ceiled. Chemical analyses of the water before and after filtration have been made by Professor Kastle, and biologically by Professor Gorman, both of the Kentucky State University.

The following additional information has been kindly given by Mr. Charles, under date of May 16, 1896:

"I wish also to make some statements regarding our filter plant, which were not provided for in your circular. To begin with, it was almost if not quite as important to deodorize or remove the smell from our water in hot weather, as to remove any solid materials or foreign matter in suspension. Our experience has been that it is usually not a difficult matter to filter water so that it will appear clear, and yet smell offensively, because if water is impregnated with a gas, the filter which passes the water will still more readily pass any gas in solution (the gas being much rarer), and if closed or what is called pressure filters are used, such gas can only escape at some point beyond the filters, generally at the outlet where consumed.

The only remedy for this smell is thorough aeration in combination with filtering, and we think we have provided for aeration to a greater extent than has yet been done, and in a simple and inexpensive manner, the details of which I

will give further on. The plan has worked successfully with us, and not one complaint has yet been made as to either the clearness, purity or smell of our water since the completion of our filter, while before that time, cattle would refuse to drink it except when compelled by necessity, and it was disagreeable to even bathe in, to say nothing about drinking it.

This result has been secured by the adoption of mechanical filters in combination with aerating devices. We could not afford filter beds, as they were much too expensive, and we did not have the space on which to locate them. As we also have no sand here, the importation and renewals necessary on the filter-bed system would have made the maintenance too expensive. We, therefore, adopted the mechanical filter (gravity) made by O. H. Jewell Filter Co., and have seen no reason yet to regret our choice. We selected the gravity filter in preference to the pressure, because it afforded greater facilities for aeration, and because the head or pressure on a gravity filter is much more uniform than with a pressure filter and its results necessarily more uniform. It is also always open to inspection.

We use a low-service pump, whose only duty is to take suction (about 5 feet) and throw (about 20 feet) to the filters. Although this involves handling all the water twice, yet we think it the best mode. The steam and water pistons are so adjusted to each other as to produce the best result and the pump works easily and continuously, and is supplied by the same boilers which are used for our large force pumps which supply the city from the clean water basin. We use direct pressure in supplying the city and our pumps are the Holly or Gaskell pattern.

The low-service pumps deliver the water at the top of the open filters through horizontal tubes or pipes spanning the filters. These tubes are 5 inches in diameter and on the under side each one is perforated with 1,000 holes of ⅛-inch diameter, through which the water falls in a spray about 7 feet on an average to the filter below. These pipes have caps which can be removed to swab out the inside, if clogged.

The roof of the filter house is high, and at the apex or summit has a sort of a dome with adjustable openings designed to create a draft and assist in carrying off any gases which escape from the water during the spraying. When the water thus sprayed has passed through the filters it is delivered to the clear water covered basin, but it only reaches the clear water basin by passing through a vat of charcoal and again falling in a spray to the basin below. This charcoal vat is 10 feet square and has a perforated zinc bottom with ⅛-inch holes all over it 1 inch between centers.

We also have an air compressor side by side with our low-service pump by means of which we can blow sprays of air through the sprays of water and can also inject air in any desired manner directly into the pipe supplying the filters, but so far there has been no necessity for its use, as the cheaper forms of aeration by spraying have been sufficient to entirely remove all odors.

While thorough aeration is important and necessary to remove the odors in water, yet it is almost equally important to get rid of the air again as quickly and thoroughly as possible after it has performed its work on account of the danger to the pipe lines from air pockets and water hammer. We accomplish this by delivering the filtered and aerated water at the surface of the clear water basin, which is 110 feet long, 40 feet wide and in which the water is 10 feet deep. The suction for the pumps which supplies the city is 100 feet horizontally distant from the point where the filtered water is delivered and 10 feet below it. Of course, air will not travel that distance or descend vertically that depth through water when opportunities are given for it to escape, and the remedy has proved sufficient with us and we have had no trouble whatever on that score.

We have facilities for injecting live steam into each filter whenever necessary for the purpose of destroying any living organisms. Our filters are washed twice each day with reverse currents and

71

thorough stirring by the agitators and the sand (white) perfectly cleaned.

I attribute much of our success to the manner in which we use our congulant (alum). We have a little simple device somewhat on the principle of an injector operated by the rock shaft of our low-service pump and working automatically, which at every stroke of the piston injects a fixed and invariable amount of the alum solution. It is only necessary to vary the strength of the alum solution from day to day or even from hour to hour if necessary, and it needs no further attention. The results are perfectly uniform; it requires much less alum and is positive and certain irrespective of variations in pressure. I figure the added cost due to filtering at $6 per 1,000,-000 gallons, which covers every expense of labor, coal and materials, but much better results should be obtained where the population justifies it, and the filtering is carried on on a large scale.

The cost of coal is an important item. Ours is hauled three miles over a toll road in carts and costs us 10 cents per bushel of soft coal, 72 pounds per bushel. If we could get our coal at 5 cents per bushel we could reduce the cost to $4.50 per 1,000,000 gallons, and with double or treble the consumption could reduce the expenses still further.

The entire cost of our plant, including buildings, clear water basins, pumps, air compressor, piping, steam engines, etc., was $27,000, and was money well expended. This is far cheaper, we think, than any filter-bed could be constructed for, and cheaper to maintain.

The results have been perfectly satisfactory, and we do not know of a thing we could change to advantage, and this statement is made after having seen and inspected some of what are generally supposed to be the best filter plants.

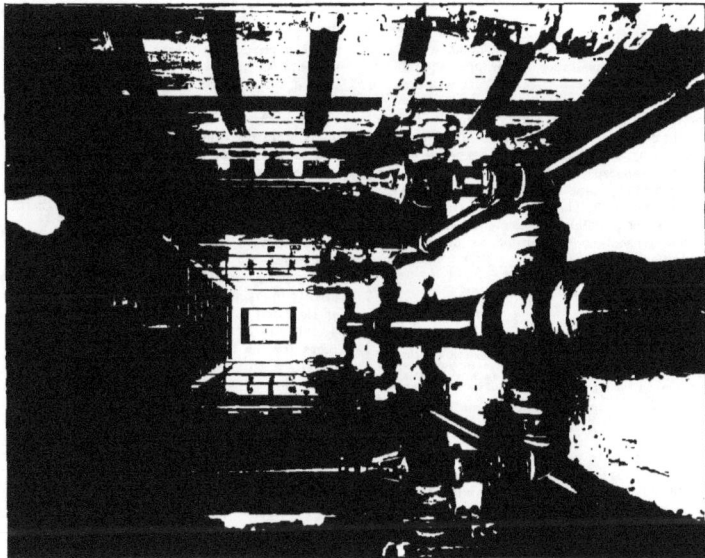

Jewell Gravity Filters at the Water Works Pumping Station, Ottumwa, Ia. Capacity, 2,000,000 Gallons Daily. Purifying Des Moines River Water. View below Gallery.

Jewell Filters at Ottumwa, Ia.

[From *Ottumwa Courier*, Dec. 10, 1895.]

THERE occurred this afternoon an event of great importance to the present and future welfare of Ottumwa. It was the formal inauguration of the gravity system of water filtration recently installed for the Ottumwa Water Works by the O. H. Jewell Filter Company, of Chicago, at an expense of $25,000. In response to the following neatly printed invitation, sent out by Superintendent Wing, some 200 of the business men of Ottumwa and a number of prominent gentlemen from other cities were present on this auspicious occasion. The invitation read as follows:

Office of
THE IOWA WATER COMPANY.
WM. A. BROWNELL,   ABRAM WING.
Receiver.   Superintendent.
FRANK E. FLANDERS, Chief Engineer.
Believing that you, as a citizen of Ottumwa, and a patron of the Iowa Water Company, are interested in the question of a good and pure water supply for the inhabitants of the city, we have thought it very proper to invite you to be present during the afternoon of Thursday November 19, to inspect and see the practical workings of the new filtering plant, at which time it will be started.
We trust that we shall be favored by your presence.
Very truly yours,
IOWA WATER COMPANY.
ABRAM WING, Supt.

The works were in apple pie order, being as neat and fresh as paint and painstaking cleanliness could make them, and the party were given opportunity to thoroughly inspect them in every detail, a privilege that was much enjoyed by all present. Comparatively few people, even in Ottumwa, have known of the important work that has been quietly but steadily going on at the water works during the past six months, and many were the expressions of surprise and pleasure on the part of those present this afternoon at what was revealed to them. They saw in actual and practical operation of what is now recognized to be the best system of water filtration in the world.

They saw the muddy Des Moines River water going into the pumps, and then witnessed the same water, after passing through the Jewell gravity filters, flowing forth in a beautiful stream, as clear and sparkling as crystal. It was a wonderful transformation and one that means volumes for the health, welfare and prosperity of the city of Ottumwa. The "Courier" believes that many thousands of its readers will be interested in a description of this simple and yet truly remarkable gravity system of filtration and to that end, we have secured the cuts appearing herewith, which will give an excellent idea of the new filters and the manner in which they are constructed and operated.

Cut on page 78 is a splendid representation of

the large filtering tanks, six of which are used in the new Ottumwa plant, with space reserved for two more which can be added as the growth of population and consequent increased use of water requires it. Cuts on pages 73 and 74 are actual views made from photographs of the Ottumwa plant. They show the interior of the new filter house, which is a substantial, fire-proof brick building, 95x41 feet. Illustration on page 74 shows the rows of tanks on either hand as seen from below the platform. The big pipe that passes between them is the main feeder from which branches lead into each tank. View on page 73 shows the rows of tanks from above, as seen from the platform, and shows the line shaft and pulleys for the agitating apparatus which is used when cleansing the filters.

A line of shafting runs the full length of the building, and to it the agitators are geared so that any one may be used independently of all others. The river water is pumped direct into the subsiding basins of the tanks by a new triplex Holly low-service pump, specially designed for the Ottumwa plant.

A COAGULANT PLANT.

In addition to the regular plant above described, and shown in the cuts, there are three additional tanks, known as the coagulant plant. This is necessary from the fact that some waters, like those of our rivers, contain more or less pulverulent and organic matters which are often so finely divided that it is quite impossible to remove all of this class of impurities by simple filtration on a large scale, no matter how judiciously the filtering-beds may be composed. In such instances and to remove bacteria, organic impurities, etc., the water is treated with lime and alumina prior to filtration for the purpose of coalescing these finely divided substances. The apparatus for this purpose, called the coagulant plant, is erected complete but will only be used at certain seasons of the year, when the condition of the water may require it.

AIR INTRODUCED IN THE WATER.

Many waters also contain, especially during certain seasons, more or less dissolved gases which impart a disagreeable odor and taste. To remove these the Jewell system introduces a copious supply of air into the water as it enters the filters, whereby the gases are displaced and eliminated. This aeration also renders the water sparkling and highly palatable.

STERILIZING THE FILTER-BEDS.

The Ottumwa plant is completely equipped with steam pipes and connections for sterilizing the filter-beds, in case this should at any time become necessary. The process requires only about an hours' time and ordinarily need only be done at long intervals.

78

THE FILTERING MATERIAL.

In all the Jewell plants the filtering material is specifically prepared for each plant on scientific formulae, the kind of material being dependent upon the character of the water to be filtered.

CAPACITY OF THE OTTUMWA PLANT.

The capacity of the Ottumwa filter plant is 2,000,000 gallons per day of twenty-four hours, and it is not intended that anything but filtered water shall hereafter be pumped into the mains. The works are built acording to the most approved plan, and in accordance with the latest discoveries of scientific investigation into the methods of water filtration. The plans were drawn by Mr. O. H. Jewell, president and manager of the Jewell Filter Company.

This company is the owner of many patents, most of which are the inventions of Mr. Jewell himself. Mr. Jewell is both a civil and mechanical engineer and has devoted twenty years to scientific investigation and practical experiments in filtration. His company is recognized authority in the United States on the subject, and the plants erected by this company are found in almost every civilized country in the world.

Recognizing the great difficulties to be overcome in successfully filtering the Des Moines river water, the Jewell people have taken special interest in the Ottumwa plant, expecting, as they have

done, to here practically demonstrate the superior results to be obtained by their system of filtration. Mr. O. H. Jewell, president of the company, drew up the plans, and, during the past six weeks, Mr. I. H. Jewell, second vice president of the company, has been in Ottumwa at various times and has personally superintended the construction, taking particular interest in seeing that everything in connection with the Ottumwa plant was done in the best manner, so as to guarantee unquestioned success. That this result has been accomplished, none of the large number of gentlemen who inspected the plant to-day can doubt for a minute. To see the clear and sparkling stream of water flowing from the filters was an inspiration, and a matter that caused the greatest enthusiasm among the spectators. It goes without saying that there is no matter of greater importance to the health and welfare of a city than a pure and wholesome water supply. It is a matter of general congratulation to be able to announce that this result has been secured for Ottumwa. It has cost many thousands of dollars, but the expenditure is such that it can not fail to receive merited appreciation and return big dividends to the investors. The "Courier" feels that Ottumwa is to be congratulated on having secured this magnificent plant, built and equipped, and now successfully installed by the O. H. Jewell Filter Company.

77

PUMPING STATION.   FILTER HOUSE.

Plant of Jewell Gravity Filters, Erected for the Cedar Rapids Water Company.
Capacity, 3,000,000 Gallons daily.   Purifying Cedar River Water.
Exterior view of Filter House and Pumping Station.

CEDAR RAPIDS, IA.

Plant of Jewell Gravity Filters, Erected for the Cedar Rapids Water Company.
Capacity, 3,000,000 Gallons daily.   Purifying Cedar River Water.
General Interior View of Filter House.

79

General Description of the Cedar Rapids Plant.

THE population of Cedar Rapids in 1800 was 18,020. A public water supply was introduced in 1875-6 by the Cedar Rapids Water Company. In 1800 the supply was reported as being drawn from artesian wells for domestic use, and the Cedar River for fire protection. Later, however, it was found necessary to use a portion of the river water, and filters were put in to clarify the water supplied from both the river and wells.

The filter house is a beautiful structure, adjoining the old pumping station. It was built much larger than necessary, to accommodate additional filters which will be put in as the consumption increases. The view on page 79 shows the space reserved for this purpose. The filters, as will be seen in above view, are arranged in two rows, with the large supply main or feeder in the center, and two wash pipes on either side of it. The filters are washed with filtered water, which is taken out of a clear water basin, located directly under the filters, by an independent low service pump. The wash water is carried off by conduits in the supporting walls, similar to the Lorain plant.

The entire cost of the filter plant, including a brick building 40x140 feet, 20 feet high, with a clear-water reservoir beneath it, two 3,000,000 gallon low-service Worthington pumps, a 5x16 foot boiler, and some land, was about $47,000.

This plant has been in operation only a few months, having been started in the spring of 1806, but, so far, no coagulant has been used or found necessary. The apparatus for its use is, however, installed complete, and ready for use whenever the condition of the water requires it.

SPRECKELS SUGAR REFINERY.

Plant of eight Steel Tank Jewell Gravity Filters at the Spreckels Sugar Refinery, Philadelphia, Pa. Capacity, 4,000,000 Gallons per day.   Purifying Delaware River Water.

BRYANT PAPER COMPANY.

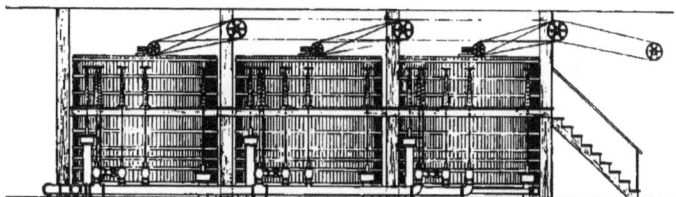

Front - Elevation

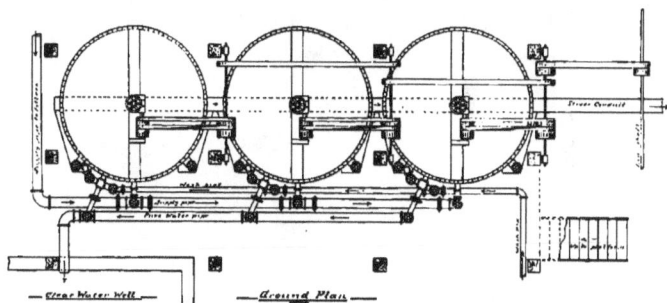

Clear Water Well — _Ground Plan_ —

Jewell Gravity Filter Plant at the works of the Bryant Paper Company, Kalamazoo, Mich.
Capacity, 1,000,000 Gallons Daily. Purifying Kalamazoo River Water.
General Plan and Elevation.

TERRE HAUTE, IND.

Jewell Gravity Filters, at the Water Works of Terre Haute, Ind.
Capacity, 1,000,000 Gallons Daily.   Purifying Wabash River Water.

83

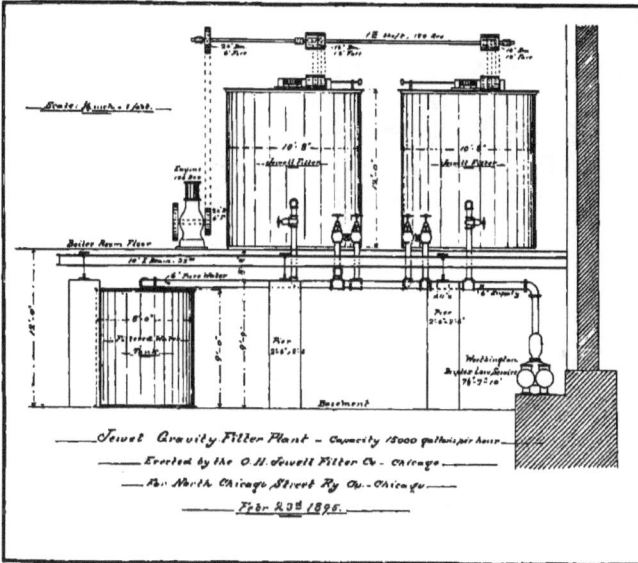

Jewell Gravity Filters, at the Hawthorne Ave. Power Station of the North
Chicago Street Railway Co.
Purifying Chicago River Water.

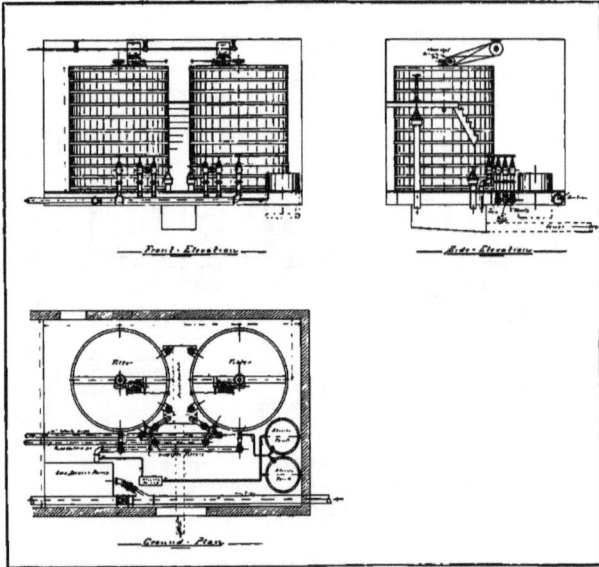

Jewell Gravity Filters at the Works of the American Glucose Co., Peoria, Ill.
Capacity, 750,000 Gallons Daily.   Purifying Illinois River Water.

DETAILS OF CONSTRUCTION.

THE accompanying views in detail of some of the parts of the modern Jewell gravity filter will illustrate the high character of the workmanship and durability which prevails throughout the entire construction.

Fig. 1 shows our improved method of fixing the screens to the parallel distributing pipes. The clamp and screws, as well as the entire screen, are of brass with the exception of the perforated metal disc and rivets, which are of bronze and copper respectively. The nozzle of the screen where it enters the pipe is slightly tapered and the screen is first driven or wedged in the pipe and then the binding screws are firmly tightened, thus making a perfectly tight joint and one which cannot possibly work loose. The many advantageous features in this plan over the former method of screwing the screens directly into the pipe, which are apparent without mention, fully compensate for the increased cost to us.

Fig. 2 represents our special washout valve, used on all our gravity filters. This valve is designed to fit directly upon the tank to which it is secured by studs or bolts. It is tapped or flanged on the bottom to receive an 8-inch pipe; the valve disc is swiveled and the seat is fitted with a hard brass removable ring. The valve stem is heavy brass, with square rapid thread. Among the many features of excellence it possesses over ordinary valves and fittings is that the large opening is longitudinal, thereby strengthening the tank by permitting the hoops to be placed close together. This valve is so constructed that no leakage, binding or sticking is possible.

Figure 3 illustrates our new automatic "controller," attached to the outlet or pure water valve of the filter. The great convenience alone of such a device is very manifest. It controls or regulates the output of the filter uniformly under the most varying conditions, saving considerable care and attention in the ordinary manipulaitons of a large plant, as well as insuring a perfectly uniform quality of the effluent. It consists essentially of a heavy cast iron chamber, with integral divisions forming a sealed outlet, upon which is mounted a perfectly balanced regulating valve which is actuated by a copper ball floating in the outlet chamber, in which the ball maintains a constant level of water, which may be readily varied to increase or decrease the delivery of the filter. The lower part of this outlet chamber is fitted with a cap containing hard brass rings of various internal diameters, also adapted to increase or diminish the rate of filtration. These rings are readily taken out or changed from the top of the controller in case of fire.

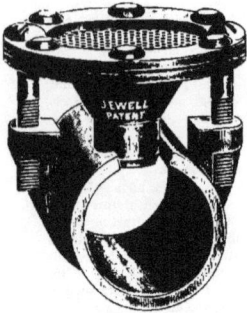

Fig. 1.
Jewell Clamped Strainer.

Fig. 2.
Jewell Tank Valve.

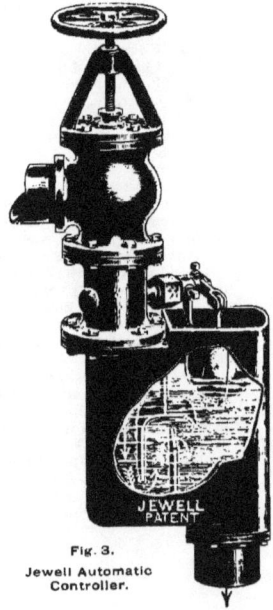

Fig. 3.
Jewell Automatic
Controller.

A Few Testimonials of Gravity Filter Plants.

Wilkes Barre, Pa., Feb. 19, 1896.
The Morison-Jewell Filtration Co., New York.

Gentlemen: In reply to yours of the 18th inst., if you will send me such a letter as you desire regarding the efficiency of the filter plant constructed by you last year for the Wilkes Barre Water Company, I will cheerfully sign it, as, for perfection of mechanical construction and simplicity of operation, I do not think you could frame a letter which would be beyond its merits.

ROGER M'GARRY,
Supt. Wilkes Barre Water Company.

Cedar Rapids, Iowa, July 6, 1896.
O. H. Jewell Filter Company.

Gentlemen: I want to say to your company that we are more than pleased with the new filter plant, and it gives us the best of satisfaction. Our citizens also appreciate the improvement, and say it is the best Cedar Rapids has ever had.

We spent considerable money investigating the several kinds of filters, and concluded the Gravity Filter would give us the best results from a mechanical standpoint, and we feel we have made no mistake. Respectfully yours,

CHARLES J. FOX, Supt.

The Home Water Company,
Royersford, Pa., June 27, 1896.
The Morison-Jewell Filtration Company, 26 South Fifteenth Street, Philadelphia, Pa.

Gentlemen: Since our filter plant was constructed by you, some three years ago, it has never failed to give our citizens a bright, clear, healthful water, even when the Schuylkill River was in its worst condition.

Our filter plant has been inspected by the Water Boards of many other cities, and has invariably been praised by them as a model plant. Yours very truly, JOHN M. SHADE, Supt.

Burlington, Iowa, July 13, 1896.
O. H. Jewell Filter Company, Chicago.

Gentlemen: In reply to yours of recent date, I am pleased to say our Jewell Gravity Filter plant of 3,500,000 gallons capacity, erected by you, has given us entire satisfaction in every particular. It has been in operation now over two years and has not cost us one cent for repairs.

In regard to cost of operating, etc., I would say it cost us last year, 1895, $3.58 per 1,000,000 gallons to filter our water. This cost includes coal for pumping, with low service pumps to filter, coagulant, light, oil, etc., and labor of two men—one night and one day.

The percentage of our pumpage in 1895 used for washing filters was 3.97 per cent. We have not, at any time during the year, washed our filters less than once every twelve hours; were it not for our direct system of pumpage for fire service, this could be reduced to washing once every twenty-four hours, certain months in the year, thereby greatly reducing the wash water, also labor.

If I can give you any further information in regard to our filter plant, will be pleased to do so. Wishing you success, believe me, yours truly,

CHARLES HOOD, Supt.

93

Office of the Board of Water Commissioners,
City Hall Building,
Danville, Pa., June 29, 1896.
The Morison-Jewell Filtration Company, 26 Cortlandt Street, New York City, N. Y.

Gentlemen: The filter plant which you constructed for us has now been in operation a sufficient length of time for us to judge intelligently as to its efficiency, and we take pleasure in testifying as to the entire satisfaction it has given us.

We have tested it severely under various conditions of our water supply, and it has fulfilled the original guarantees you made.

We are fully satisfied with our investment, and feel that we have the best system of filtration which has yet been designed.

Yours very truly,
R. K. POLK,
Chairman Board of Water Commissioners.

Ottumwa, Iowa, March 30, 1896.
O. H. Jewell Filter Co., Chicago, Ill.

Gentlemen: In regard to the Gravity Filtering plant recently erected for this company by you, will say that it is a success in every particular. The Des Moines River water has been about as muddy and filthy as it could be under any circumstances, but the filters are doing their work in grand shape, the dirty, muddy water coming out of the filter as clear as crystal.

At the time the advisability of our putting in a system of filters was considered, the question came up, what one will we adopt? And after looking over the various kinds on the market, it was decided to put in yours, and we are pleased with our decision, and think if any one is thinking of putting in a system of filters they will make no mistake by adopting the O. H. Jewell Gravity Filters. Yours truly,
ABRAM WING, Superintendent.
FRANK E. FLANDERS, Engineer.

STATE OF IOWA, WAPELLO COUNTY, ss.:

Abram Wing, being duly sworn, says that he is a resident of the City of Ottumwa, County of Wapello, State of Iowa, and that since Aug. 4, 1894, he has been general manager and superintendent of the Iowa Water Company, of the said City of Ottumwa, and he further says that, in June, 1895, William A. Brownell, the receiver of the said Iowa Water Company, having been duly appointed by Judge Woolson, of the United States Court, in chambers at Keokuk, County of Lee, State of Iowa, contracted with the O. H. Jewell Filter Company, of the City of Chicago, County of Cook, State of Illinois, to construct for the Iowa Water Company a system of what is known as the O. H. Jewell Open Gravity Filters, said filters to have a capacity of 2,000,000 gallons each twenty-four hours; said filters to be of their latest improved, and to have what is known as the subsiding tank or basin, which is directly under the filter-bed. The said filter plant has been in successful operation since Dec. 19, 1895; and he further says that since Dec. 19, 1895, the subsiding tank or basin had been opened at intervals of from thirty to forty days, and at each time there has been collected in the said subsiding tanks or basins solid mud to the depth from twelve to six inches, and this was after the valve to the subsiding tank or basin had been opened so as to allow the water to be discharged, which, of course, carried off a large quantity of slush mud to the sewer. We are fully of the belief that the sub-

89

siding tank or basin retains at least 50 per cent. of the impurities in the water before passing to the filter bed.

ABRAM WING.
Subscribed and sworn to before me, by the said Abram Wing, this 22d day of July, 1896.
O. J. GARRIOTT,
Deputy Clerk District Court.

STATE OF IOWA, WAPELLO COUNTY, ss.:
I, Frank E. Flanders, on my oath, depose and say that I am a resident of the City of Ottumwa, County of Wapello, State of Iowa; that I have been chief engineer of the Iowa Water Company during the time the O. H. Jewell Open Gravity Filters have been in operation for said company; that I have read the foregoing affidavit of Abram Wing, and that the statements therein contained are true, as I verily believe.

FRANK E. FLANDERS.
Subscribed and sworn to before me, by the said Frank E. Flanders, this 22d day of July, 1896.
O. J. GARRIOTT,
Deputy Clerk District Court.

Lexington, Ky., Jan. 23, 1896.
O. H. Jewell Filter Company, 73 and 75 West Jackson Street, Chicago, Ill.

I take pleasure in stating that the filter plant you built for us has more than filled our expectations, and is perfectly satisfactory. It does what no other filter plant I know of does, in a simple and efficient manner—it thoroughly aerates and deodorizes the water and removes all smell, a very important matter, especially in warm climates where vegetation is rank.

Every guarantee you have made has been more than fulfilled, and a number of improvements and betterments have been made by you, without charge, which were not contemplated in our contract.

We think we have the best (not the largest, of course,) filter plant in this country, and we cheerfully invite comparison with any other.

Wishing you all the success that your work so well deserves, I am, very respectfully,
S. A. CHARLES, Supt.

Waterloo Water Company,
Waterloo, Iowa, July 28, 1896.
O. H. Jewell Filter Company, Chicago, Ill.

Gentlemen: In reply to yours of recent date, making inquiry in regard to how we like the Jewell Filters for six years. We tried several other plans for getting pure water without success. One was a filter-bed built along the river bank. This was a failure. Then we tried filtering through eighteen inches of sand, held upright by two sets of fine screens. This also was a failure. Then we tried a filter with two sets of valves, one for filtering and the other for washing, this also was a failure.

Six years ago we put in a 10-foot Jewell Gravity Filter, costing us about $2,000 for the filter alone. This proved satisfactory, but in two years the increased consumption of water required more filtering capacity, and we put in a similar 10-foot filter. These two have furnished us all the water used until a few months ago, when we put in another Jewell Gravity Filter of the same size, but of the improved pattern, with internal subsidence basin, which we find has a much larger capacity than the former styles.

We like the Jewell Filters very much. They do the work well, and are easily handled and eco-

95

nomical. Of course, we have to pump the water twice, but we think it is cheaper to do this than to put a pressure filter between the pumps and the consumer, as the expense caused by increased pressure will be more than the cost of pumping it twice, as we work against the small head with the low-service pump, and the cost of this pumping is very little.

We use aluminum sulphate when the water is roily, but a good deal of the time we do not use any. I will not go into details, but, as you know, people don't buy that which, after being tested, does not give satisfaction, and the fact that we have been making an additional purchase is enough to convince anyone that the Jewell Filter has been a complete success.

As to the additional cost of filtering the water, I am not in position to state, but think it could not exceed 50 cents per 100,000 gallons.

Your new filter is a great improvement, and I

do not see how it could be bettered in any way. We now wish they were all of the new design. You may refer any parties to us whenever you wish. Very truly yours,   J. P. BERRY,
Superintendent W. W. Co.

———

Spreckles Sugar Refining Company,
Philadelphia, Pa., June 29, 1896.
To whom it may concern:
The Morison-Jewell Filtration Company, over two years ago, erected eight (8) Gravity Filters at this refinery, aggregating four million (4,000,000) gallons capacity in twenty-four (24) hours. It gives us pleasure to state that this plant has given us entire satisfaction, and that it more than fulfills the guarantees made by the Morison-Jewell Filtration Company. Yours very truly,
CHARLES WATSON,
Chief Engineer Spreckles Sugar Refining Co.

The Jewell Pressure
Filter.

Standard Vertical Type,
with Subsidence
Basin.

These Filters are espe-
cially adapted for public
and private institutions
of any magnitude, office
buildings, hotels, clubs,
natatoriums, apartments,
breweries, bottling works,
chemical works, laund-
ries, etc.

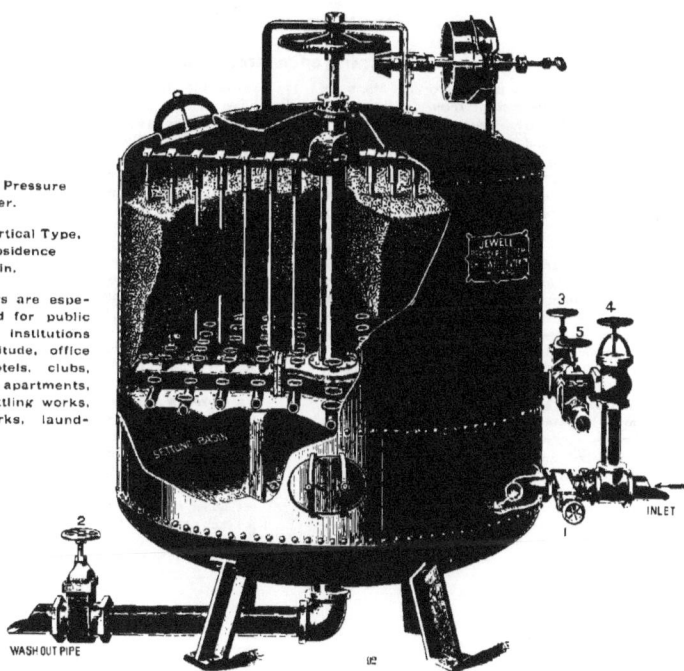

JEWELL

SETTLING BASIN

3 4
5

INLET

2

WASH OUT PIPE

97

Jewell Pressure Filter.

DETAILS OF CONSTRUCTION.

ALL of our standard pressure filters are made in vertical form; the heads are made extra heavy and are dished to withstand heavy internal pressures without any bracing whatever. All seams are double riveted (on the large sizes) and the shells are mounted upon heavy cast iron supports, which present a neat and durable appearance and render all parts and connections easily accessible. The man-holes, when admissible, are placed in the top head, and are of the well known Eclipse pattern. On the small sizes the man-holes are placed on the side in such a position as to enable one to reach all parts from the outside. The large sizes are also provided with man-holes on the lower side, opening into the subsidence basin, to facilitate riveting and caulking of the seams. The inner bottom or division plate is extra heavy, and in addition is supported from below by heavy wrought iron stands. There is no strain or pressure either way on this plate except the weight of the filter-bed and wash system, as the central standpipe affords perfect equalization of pressure. The wash and collecting strainer system is fastened directly to this plate, and a heavy, threaded, wrought iron flange, riveted to the center, receives the standpipe, which rises several inches above the level of the filtering-bed. The wash and collecting system is substantially the same as in the gravity filters, and consists of several heavy cast iron manifold sections laid diametrically across the inner plate. These are securely bolted together, and connected at the front end with a short pipe leading directly to the valves 3, 4 and 5. (See page 92). The manifold sections above referred to are provided with a series of threaded openings along either side, into which are screwed the extra heavy wrought iron branch pipes; and into these pipes, every few inches, are clamped the numerous strainers. (The strainer and method of securing same are shown on page 87). There are several hundred of these brass and bronze strainers in each filter of the larger sizes, and the combined area of the nozzles of all is somewhat less than the area of the supply pipe, although the combined area of the bronze screen itself is over ten times the area of the pipe or nozzles. The object of restricting the area of the nozzles is to compel a uniform distribution of water and pressure under the bed during cleansing and to compel each part of the filter-bed to perform its proper

98

functions. This feature is patent only to the Jewell filters and is, of the many points of superiority, one of greatest importance. The filter-beds are all from three to four feet in depth, and composed, in most instances, of white machine-crushed quartz, though we also use, in some instances, beds of extra-sifted white silica, which is most uniform in size and quality, and, for permanency, can not be equaled. The bars or rakes of the stirring apparatus penetrate the beds almost to the strainers themselves, and are located on the horizontal shafts so as to positively scour every portion of the bed alike. The horizontal shafts, upon which the rakes are secured, are held by a heavy cast iron cross-arm, which revolves with the vertical shaft, passing up through the stuffing box to the bevel gear on top of the filter. All of these parts are carefully fitted, and the workmanship or construction can not be excelled. The saddle which combines the gear and pinion is wrought iron and very heavy. The guide in the stuffing box is brass and the vertical shaft is steel encased, with drawn brass tube, so that no corrosion or sticking of the working parts is possible. The pinion shaft is fitted with tight and loose or clutch pulleys, as preferred. The gears, cross-arm, etc., are all keyed to the shaft, and in addition are held by one or two set screws. All valves used on the filters are of the double-wedge gate type, which are either entirely of brass or brass mounted. The pipes and fittings are all standard and first-class in every respect. There is, in fact, no feature of advantage at present known in the art of filtration which has been omitted in the construction of these filters. They are made for steady and severe service, and for durability, efficiency, and economy can not be excelled or even equaled, in any particular. These filters are not merely huge strainers, built as cheaply as possible, to sell and shortly be abandoned, but are effectual purifiers of the water, rendering the same bright and crystalline during all conditions of the supply, and will work constantly and uniformly without any repairs or labor for many years.

DIRECTIONS FOR CONNECTING.

The supply or pipe for bringing the natural water to the filter is connected at the T-marked "inlet," which branches to valves 1 and 4.

The delivery, or pipe for conveying off the purified water, is connected to valve 5.

The waste, or pipe for carrying off the dirty wash water, is connected to valve 2.

The re-wash, or pipe for discharging to waste the first water filtered after washing, is connected to valve 3. This is used when the filter is washed with the natural water in order to displace the turbid or impure water in the filter, or for draining the filter.

All pipes, except pure water pipe, which may be considerably smaller, should be the same size as the valves to which connection is made. The filters should be placed on level and durable foundations. The proper speed to drive the agitators is from six to eight revolutions per minute, preferably the former figure. If the agitator is run too fast it may result in a partial loss of the filtering material.

METHOD OF OPERATION.
WASHING.

To wash the filter, all valves having been previously closed, open waste valve 2 full; this relieves the pressure on the filter, drains water entirely out of settling basin and lowers water level in filter to level of stand-pipe. Then open valve 4 slowly until the agitator revolves easily; a little experience will enable the operator to adjust it very readily. If the pressure on the wash pipes is not constant, a pressure guage inserted in the plugged hole on top of cross will be a great convenience and insure uniform results. The water pressure during washing should not be less than 5 pounds at that point; and, on the other hand, if too much, may result in a partial or gradual loss of the filtering material. Do not start agitator until wash water is circulating freely, and always stop it before shutting off the water, as the bars or rakes will not pass through the bed unless same is in a semi-fluid condition. The water used in washing enters through valve 4 and passes directly into the mani-

fold distribution system; thence issues upward through the numerous strainers like a myriad of geysers, which causes a violent ebullition or commotion in the bed, thoroughly liquifying same, uniformly in all parts. The upward current of water carries with it the released impurities and flows freely over and into the central stand-pipe, from whence it is discharged into the settling basin and is sprayed to all parts of same by a deflecting plate and then discharged through opening in bottom to sewer, thereby thoroughly cleansing the filter-bed and settling basin in one operation, without any additional expense for water or labor. When washing is complete and the filter clean, first close valve 4, then close valve 2. This operation should not require over ten minutes' time.

If the filter discharges the purified water directly into a standpipe or elevated reservoir, it may be washed with the back pressure by simply opening valve 5 to the required extent instead of valve 4, in which case the latter valve would not be used at all. It is advisable to wash with filtered water wherever practical. The amount required is a very small percentage of the amount filtered.

RE-WASHING.

After washing the filter, open valve 1 (all valves having been closed) and when filter is re-filled and under working pressure, open valve 3 slowly one to two turns only. Then sample the water from try cock on bottom of cross, and when same becomes clear and satisfactory in color and appearance, close valve 3, allowing valve 1 to remain full open. After cleansing the filter, especially if the filter is washed with the natural or turbid water, it is customary to waste the first water filtered after washing, in order to prevent same entering the clear water or service pipes; the idea being to simply displace the muddy water remaining in the lower parts of the bed by clean, pure, filtered water, and, therefore, this operation only requires a moment or two and may be discontinued when the effluent becomes clear. For this purpose, a try cock is put in the bottom of the "cross" to sample the filtered product. When filtered or purified water is used for washing the filter, it is seldom, if ever, necessary to re-wash, as above described. In the operation of re-washing, the turbid or natural water enters the settling basin and eventually passes up the central stand-pipe, thence flows onto the filtering bed, percolates downward through same into the strainer system, and is discharged through valve 3 to the sewer, and is, therefore, almost identical with filtering.

FILTERING.

After the filter has been re-washed, as above, and all valves closed except valve 1, then open valve 5 slowly, in order not to cause a sudden heavy draft on the filter, and when pressure in pipes is equalized, this valve may be fully opened. The filter is then in full service. In this operation the water enters the settling basin same as in re-washing, depositing therein all of the heavy impurities before the water is discharged from it to the filter-bed. The alumina is completely precipitated as a flocculent mass and the impurities so thoroughly coalesced that even bacteria and extremely fine silt or clay particles are arrested in this basin to the extent of 30 to 40 per cent., in addition to the deposition of the heavier impurities which are arrested in the basin without the aid of congulation. The water leaving the settling basin passes upward through the central stand-pipe, thence on to the filtering bed, down through which it slowly percolates, depositing the remaining impurities which have not subsided in the basin, in its interstices, from whence it is collected by the strainers, branch-pipe and manifold system, and discharged through valve 5 as clear and sparkling as the best spring waters.

Ordinarily it is advisable to wash or cleanse the filter once in twenty-four hours. On some waters, however, and during severe conditions or rises, it is necessary to wash a little oftener; and, on the other hand, many waters do not require the filter to be washed oftener than once in three or four days. This depends entirely upon the condition of the natural water, and may be given as a fair average at from once to twice a day of twenty-four hours. Of course, if a filter is pushed beyond its rated capacity, the bed becomes clogged sooner and requires washing more frequently. In practice, a filter is generally washed only when the capacity of same is below the required amount. The condition of the filter-bed, or amount of accumulation in same, may be noted by comparing the pressure before and after filtration by attaching a gauge to the inlet and also one to the outlet of the filter. By noting the maximum loss of head, as shown by the two gauges, or by noting the minimum working pressure on outlet gauge, the time for washing can be fixed very accurately.

THE JEWELL HORIZONTAL PRESSURE FILTER.
(WITH SUBSIDING BASIN.)

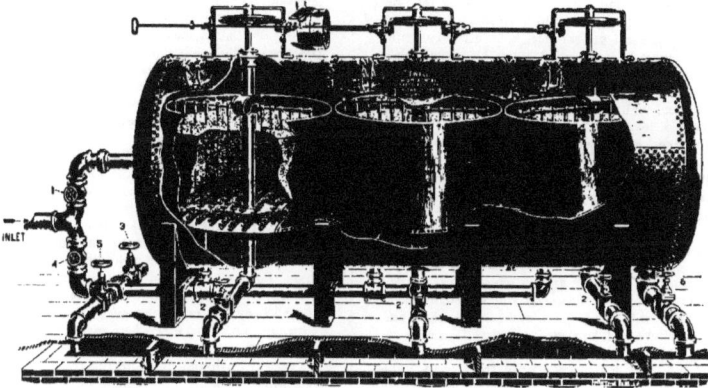

Fig. 4.

Specially adapted for High Pressure Service.

UNQUESTIONABLY THE BEST HORIZONTAL FILTER EVER PRODUCED.

The Jewell Horizontal Filter.

T will be noted by the illustration on the preceding page, this filter, although horizontal in general appearance, is in all substantial respects constructed upon the same plan as our standard vertical filters. The filtering compartments, of which there are three or more, are our usual cylindrical vertical type, and placed within a large horizontal shell, which allows a very large space for subsidence of the water prior to entering the filter chambers.

The filtering compartments are carried or mounted on the lower side of the horizontal shell by several heavy lugs, and in addition thereto are supported at the center by large tubular cast iron columns, all being substantially riveted together. Each of the filtering compartments is fitted with our complete agitating apparatus, which is carried by wrought iron saddles on the horizontal shell and fitted with bevel gears and pinions, tight and loose pulleys, shifters, etc., complete.

All pipe connections enter on the bottom, except the inlet, which enters the front head. The filters can, therefore, be completely drained to prevent freezing, and placed close together in battery form, and all connections are rendered easily accessible by the heavy cast iron or masonry stands upon which the entire filter is mounted, thus presenting a most neat and substantial appearance. The shell is provided with three manholes, one over each filtering compartment.

The filtering compartments are arranged to be washed simultaneously, but can be washed separately, by a few slight changes in the agitator and pipe system, if desired. When the available quantity of water for washing is small, as compared with the capacity of the filter, this feature of independent compartment washing will be found very serviceable.

This filter is operated in the same manner as the vertical filters, with the exception that the filtering-beds are washed independently of the settling basin, but both may be washed simultaneously whenever desired, and with no more attention or labor. They are especially adapted for locations with high pressures and where large quantities of water are used.

Every feature and detail in the construction of this filter is as near perfection as our long and practical experience could devise, and the material and workmanship are all first-class.

The Jewell
Pressure Filter.

Small Size Pattern, with
Settling Basin.

100

The Jewell Pressure Filter.

SMALL SIZE.

THE accompanying cut, Fig. 5, illustrates our subsiding tank pressure filter, as made in sizes from 12 inches to 24 inches in diameter inclusive. Larger sizes of this pattern made if desired.

The filter shells are made throughout of heavy flanged steel plate, with the exception of the lower head or base, which is heavy cast iron. This base is composed of two parts, which are firmly bolted together. One of these parts is a double flanged ring, to which the outer and inner tanks are riveted, and this part or ring is provided with three heavy legs or supports, upon which the entire filter rests. The other or removable portion of this base consists of a hollow circular chamber, into the upper side of which the wash and collecting screens are screwed. This chamber is also provided with a threaded boss, to which is connected the pipe leading to the cross and valves 3, 4, and 5. The pipe connections are substantially the same as on the larger sizes, and are simply reversed, bringing the pure water discharge below the inlet, instead of above same. For method of connecting and operating, see pages 94, 95, 96 and 97.

This filter embodies practically all of the advantages of the other pressure styles, and leaves nothing to be desired. It is constructed throughout in a first-class manner, with a view of obtaining the greatest durability, efficiency, convenience and neatness of appearance possible.

THESE FILTERS ARE ESPECIALLY ADAPTED
· · FOR USE IN · ·

Office Buildings, Public Schools, Colleges, Laundries, Bottling Works, Ice Plants, Apartments, and Residences.

TABLE OF STANDARD SIZES

OF THE

JEWELL PRESSURE FILTERS,

With Combined Settling Basin.

REGULAR VERTICAL TYPE.

Illustrated on page 92.

No.	Size.	Connections.		Capacity.			Area.	Height	Space.	Bed.	Weight. (Approximate.)			Test.
Serial.	Nominal.			Minimum and Maximum U. S. Gallons.							Shipping.			
For Convenience in Telegraphing.	Diameter.	Supply and Discharge Pipes.	Washout Pipes.	Minute.	Hour.	Day: 24 Hours.	Effective Filtering Surface.	Extreme Height over all. About	Approximate Floor Room required.	Actual Quantity of Filtering Material.	Filter, Connections, etc.	Filtering Material.	Gross Weight of Filter in operation.	Pressure. Lbs. per square inch.
	Ins.	Ins.	Ins.				Sq. Ft.	Feet.	Feet.	Cu. Ft.	Lbs.	Lbs.	Tons.	
9	12	¾	1	1¼—2½	75—150	1,800—3,600	¾	6	1¼×2	2	300	150	½	75
10	18	1	1¼	3¾—7	225—450	5,000—10,000	1¾	6	2×3¼	4	550	350	1	100
11	24	1¼	1½	5—10	300—600	7,000—14,000	3	6	2¼×3	7½	800	600	1½	100
12	36	1½	2	11—22	700—1,400	16,000—32,000	7	8	3×4	17	1,900	1,400	3	100
13	48	2	2½	20—40	1,200—2,400	28,000—56,000	12	9	4×5	30	2,500	2,500	5	100
14	60	2½	3	30—60	1,900—3,800	45,000—90,000	19	9	5×6	48	4,000	4,000	7	100
15	72	3	4	45—90	2,800—5,600	65,000—130,000	28	9	6×7½	68	5,000	5,600	10	100
16	84	4	5	63—126	3,800—7,600	90,000—180,000	38	9	7½×9	94	7,000	7,800	15	100
17	96	4	6	83—166	5,000—10,000	120,000—240,000	50	9	9×10	120	9,000	10,000	20	100
18	120	5	6	120—240	7,800—15,600	180,000—360,000	78	9	11×12	192	12,000	16,000	35	100

SPECIAL HORIZONTAL TYPE.

Illustrated on page 98.

	Feet.														
19	8×20	5	3-6	170—340	10,400—20,800	225,000 450,000	2,600	101	12	10×20	250	24,000	25,000	35	100
20	8×20	5	3—6	188—376	11,300—22,600	250,000—500,000	3,000	113	12	10×26	275	25,000	27,500	40	100

Special Sizes made if desired. Filters built to withstand up to 200 lbs. test pressure when necessary.

Apparatus for Feeding Coagulants or other Reagents.

MONG the many available ways of introducing a coagulant or reagent into the water to be filtered or purified, we have found but three that gave complete satisfaction in all respects.

1st. The use of a small auxiliary steam or mechanical pump, automatically operated by the main supply pumps.

2nd. An independent pressure tank containing the alum, connected with the supply main, having an automatic resistence valve, causing a proportionate shunt current of water to pass through the tank.

3rd. A meter directly actuating a small auxiliary pump.

These methods embody the three fundamental and most improved features, viz., proportional accuracy in feeding under the most varying conditions; positive action; and absolutely no liability of an excess of the reagent. We have made this subject one of extensive investigation and improvement; and, after many years of experience, believe we have the most reliable methods known.

AUTOMATIC AUXILIARY PUMPS.

The first of these methods is wholly confined to use on large plants, and as the conditions attending such installations require specific apparatus, we have not illustrated any particular form of same, and, in fact, the method is so apparent that no detailed description is necessary.

THE JEWELL PRESSURE COAGULANT TANK.

The second method is shown by cut No. 6, which illustrates our independent pressure coagulating tank. The alum is put in through handhole 1, cock 2 being left open while filling, to carry off waste water. Valve 4 is the regulating valve, and valves 5 and 6 are used only in re-filling, to cut off the water pressure from main supply pipe. Tri-cock 3 is opened only while regulating the alum feed by valve 4; and it indicates, when valve 4 is partially opened and valves 5 and 6 closed, the amount of alum solution being used on the water. The most important working part in the construction of this apparatus is the "automatic resistence valve" on the supply pipe. On its action depends practically the whole working of the tank, and it is therefore essential that it be constructed on the right principles. We have made a large variety of automatic valves for this purpose, and are confident that we now have one that cannot be excelled.

THE JEWELL PUMPING METER.

Illustration No. 7 represents our automatic coagulant feeding apparatus operated by meter.

through which all of the water passes after leaving the filter. The meter is preferably placed on the delivery pipes, so as to prevent cutting or stoppage by sediment in the natural water. The alumina solution is prepared in a small open tank on the right, by thoroughly dissolving a definite amount in the tank, from which a pipe is led to the suction of the pumps C, and the solution, after passing these pumps, is forced directly into the supply pipe of the filter. The pumps are bronze and have outside packed plungers. Their capacity is one-half of one per cent. of the amount filtered; thereby allowing weak solutions to be used if desired. By varying the strength of the alumina solution, any amount of alum can be fed with positive accuracy into the water. Either of the above appliances may be placed close to the filters, as they are provided with large subsidence basins; or, they may be located at any convenient place on the supply or discharge pipes respectively.

The proper kind of alum to use in the apparatus shown in cut No. 6 is commercially known as "cracked potash alum," the crystals of which are about the size of a hickory nut, although larger crystals or lumps of other alums may be used. In the form shown in view No. 7, no special kind or form of alum is required, as it all goes directly or completely into solution. The commercial sulphate of alumina is however to be preferred on account of its greater solubility, strength, and lesser cost. As a general rule, when operating filters at full capacity, it may be stated, that the amount of alum or other reagent required varies from 1-10 to 2 grains per gallon of water filtered; or, from one pound to 70,000 gallons to one pound to 3,000 gallons, depending on the character of the natural water; also, within certain limits, the amount of alum required is inversely proportional to the rate of filtration.

Fig. 6.

Fig. 7.

Jewell Pumping Meter.

The Jewell Automatic Lime Tank.

THE excellent results we have obtained by the use of lime in softening waters, and in connection with alum or sulphate of alumina for insuring proper coagulation, led to the necessity for making a definite form of apparatus for use, which is herewith illustrated in Fig. 8.

The sparing solubility of lime (about 1 part in 700 of water) renders it necessary to use either very large vats or an automatic self-feeding device, which dissolves the lime only as fast as it is used. This latter method is preferable for several reasons; viz., it is much less expensive; requires but little space; requires no manual labor or attention, and is always ready for use.

Referring to the illustration, it will be noted that the inlet is at the bottom and the outlet at the top of the tank. The incoming water (which may be under heavy pressure) is distributed evenly over the bottom by the branch pipes and strainers, and then passes upward through the crushed quartz and layer of undissolved lime lying thereon; thence on upward to the outlet at the top. The water in its slow passage forms almost a saturated solution with the lime, and the large space above permits of very long subsidence; so that the solution leaving the tank is quite clear, depending, of course, upon the rate at which the tank is operated.

DIRECTIONS FOR OPERATING.

The proper proportion of lime to aluminum sulphate to be used is 158 parts to 648 parts by weight of each respectively. Lime may be put in tank without slacking.

Add slowly each day the required amount. Regulate amount used by valve on the outlet or suction pipe. Supply the tank with clean water if possible. Once each year, or oftener if necessary, it is advisable to drain the tank and clear out the insoluble matter left on the quartz.

These tanks have given excellent satisfaction wherever used, and we believe should be used on almost all purifying plants operating on river or creek waters subject to freshets, at which time the water does not usually contain sufficient carbonates of lime, magnesia, etc., to thoroughly precipitate the alumina.

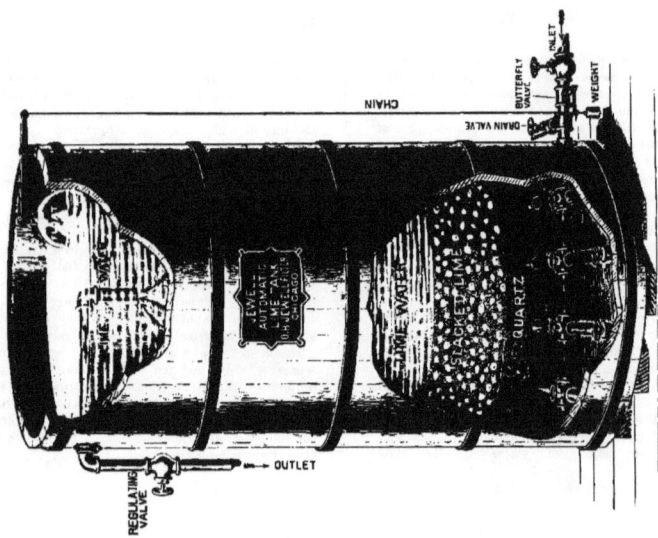

Fig. 8.

107

Description of Vacuum Chemical Tank.

Fig. 9.

THE apparatus illustrated herewith is specially designed for use on moderately large filter plants supplied by a pump, or on plants where it is inconvenient to locate large Cypress tanks in which standard solutions of a given strength are made. In this vacuum tank alumina sulphate or alum is only dissolved as fast as it is used, the tank being filled partially full with crystals or broken pieces of the reagent. The tank is kept at a uniform level by the use of a float valve on the inlet pipe, which is preferably connected to the filtered water pipes or pressure main. The outlet of the tank is taken a few inches below the top, and provided with a needle regulating valve; and this outlet may be connected directly with the suction of the main pumps or the small auxiliary pumps, the latter method being preferred. The tank is made of heavy galvanized iron. These are made in various sizes, ranging from 12 inches to 3 feet in diameter, and from 18 inches to 4 feet in height.

The Jewell Basket Strainer.

FOR PUMP SUCTIONS, ETC.

THE Jewell Basket Strainer, illustrated herewith, is made to connect directly upon a line of pipe as a trap, or used close to or upon either the suction or discharge of a pump, thereby doing away with the necessity of a strainer under water. This is also applicable for use in many other places, where it is desired to remove any coarsely suspended substances from waters or other liquids.

It is termed a Basket Strainer, in that it contains a basket through which all the water passes in the direction as indicated by arrows, and into which the impurities are retained. This Basket may be easily taken out and cleaned by simply removing the cover plate as shown in cut. The plug at the bottom being used only for blowing off any impurities clinging around the outer casing.

These Baskets are all made of No. 20 gauge copper metal with perforations 3-32 inch in diameter, having an area of more than five times that of the inlet pipe, thus allowing a large amount of sediment to be retained in the Basket without materially affecting the flow of water through the same, and under ordinary conditions only removed at long intervals of time.

The principal advantages of this Strainer, are not only to allow of its convenient application, as well as to admit of the easy manner in which impurities may be removed, but to also insure a positive and reliable method of arresting the suspended impurities. We carry these Strainers in stock with screw ends in sizes from $\frac{3}{4}$ inch to 3 inch inclusive, and in dimensions for $3\frac{1}{2}$ to 12 inch pipe connections with flange or screw ends as desired.

We guarantee these Strainers to withstand 250 pounds per square inch.

Larger sizes made if desired. Baskets made with finer perforations if desired.

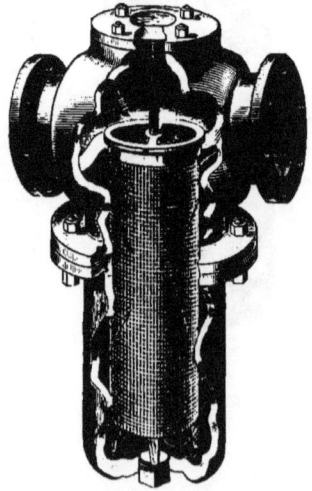

100

CHATTANOOGA, TENN.

Jewell Pressure Filter Plant at Chattanooga, Tenn., Water Works.
Capacity, 3,000,000 Gallons Daily. Purifying Tennessee River Water.

110

LAKE FOREST, ILL.

Jewell Pressure Filters, at the Lake Forest, Ill., Water Works.
Capacity, 750,000 Gallons Daily.   Purifying Lake Michigan Water.

Jewell Pressure Filter Plant at Reading Terminal Passenger Station. Philadelphia, Pa.
Capacity, 500,000 Gallons per Day.

PHILADELPHIA, PA.

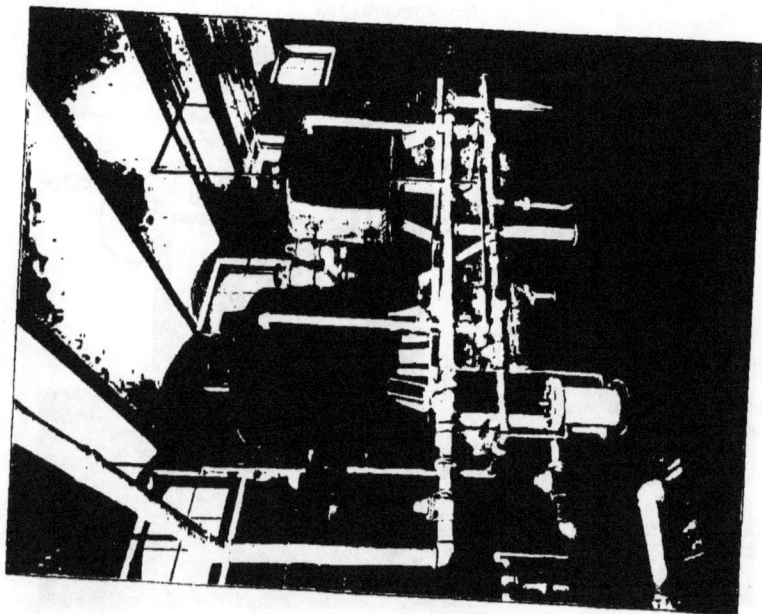

Jewell Pressure Filter Plant at Pennsylvania Hospital, Philadelphia, Pa.
Capacity, 250,000 Gallons per day.

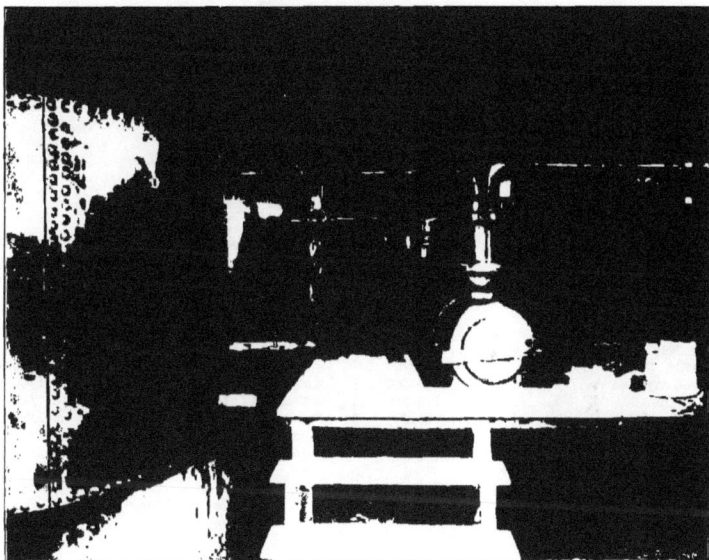

Plant of four 10-ft. Jewell Pressure Filters, at the Works of the Home Water Co., Little Rock, Ark.
Supplementing a Battery of other Filters.

114

Partial List of Filtering Plants.

CITY WATER WORKS.

LOCATION.	DAILY CAPACITY. GALLONS.
Wilkes Barre, Pa	14,000,000
Elmira, N. Y	6,000,000
Niagara Falls, N. Y	4,500,000
Quincy, Ill	4,000,000
Burlington, Ia	3,500,000
Columbia, S. C	3,000,000
Chattanooga, Tenn	3,000,000
Lorain, Ohio	3,000,000
Cedar Rapids, Ia	2,500,000
Lexington, Ky	2,000,000
Ottumwa, Ia	2,000,000
Rock Island, Ill	2,000,000
Carlisle, Pa	1,500,000
Royersford, Pa	1,000,000
Waterloo, Ia	1,000,000
Terre Haute, Ind	1,000,000
Danville, Pa	1,000,000
Merrill, Wis	1,000,000
Winchester, Ky	750,000
Lake Forest, Ill	750,000
Macon, Ga	500,000
Deseronto, Can	500,000

SUGAR REFINERIES.

LOCATION.	DAILY CAPACITY. GALLONS.
Spreckles Sugar Refining Co., Philadelphia	5,000,000
McCahn Sugar Refining Co., Philadelphia	2,000,000
Franklin Sugar Refining Co	1,000,000
American Glucose Co., Peoria, Ill	750,000

PAPER MILLS.

American Wood Paper Co., Manyunk, Pa.	3,000,000
Bardeen Paper Co., Otsego, Mich	3,000,000
Nekoosa Paper Co., Nekoosa, Wis	3,000,000
Port Edwards Paper Co., Port Edwards, Wis	3,000,000
Chelsea Paper Co., Norwich, Conn	2,000,000
American Wood Paper Co., Spring City, Pa	1,000,000
Kalamazoo Paper Co., Kalamazoo, Mich	1,000,000
Columbian Paper Co., Buena Vista, Va	1,000,000
Plover Paper Co., Plover, Wis	1,000,000
Albemarle Paper Co., Richmond, Va	1,000,000
Bryant Paper Co., Kalamazoo, Mich	1,000,000

113

LOCATION.	DAILY CAPACITY. GALLONS.
John Dickenson & Co., Ltd., Watford, England	800,000
Badger Paper Co., Kaukauna, Wis	500,000
Mt. Tom Sulphite Pulp Co., Mt. Tom, Mass.	300,000
Darblay Pere et Fils, Essonne, France	175,000

MILLS AND FACTORIES.

LOCATION.	DAILY CAPACITY. GALLONS.
Providence Dyeing, Bleaching and Calendering Co., Providence, R. I.	1,000,000
Michigan Carbon Works, Detroit, Mich.	2,000,000
Armour Packing and Provision Co., Kansas City, Mo.	1,000,000
Trenton Iron Co., Trenton, N. J.	500,000
White Bros. & Co., Lowell, Mass.	1,000,000
St. Louis Union Stock Yards Co., St. Louis, Mo.	300,000
East St. Louis Packing and Provision Co., East St. Louis, Ill.	300,000
Arlington Mills, Lawrence, Mass.	300,000
Western Tube Co., Kewanee, Ill.	500,000
H. Waterbury & Sons, Oriskany, N. Y.	200,000
Armour Grain Elevators, Chicago, Ill.	250,000
Troy Laundry Machinery Co., Chicago, Ill.	200,000
Secaucus Iron Co., Secaucus, N. J.	175,000
Wm. A. Slater Cotton Mills, Jewett City, Conn.	175,000
Fisher Bros., Detroit, Mich.	200,000
New York Dredging Co., New York City	100,000

LOCATION.	DAILY CAPACITY. GALLONS.
Ogden Gas Co., Chicago, Ill.	100,000
Barstow Thread Co., Providence, R. I.	175,000
Pullman Palace Car Co., Pullman, Ill.	160,000
Diamond Ice Co., Wilmington, Del.	150,000
Star and Crescent Milling Co., Chicago, Ill.	60,000
Hyde Park Brewery, St. Louis, Mo.	60,000
St. Louis Brewery Association, St. Louis, Mo.	60,000
Schoenhofen Brewing Co., Chicago, Ill.	60,000
Ballard Vale Spring Water Co., Ballard Vale, Mass.	50,000
E. D. Onion Ice Mfg. and Cold Storage Co., Baltimore, Md.	50,000
International Navigation Co., Jersey City, N. J.	45,000
Kansas City Ice and Coal Co., Kansas City, Mo.	45,000
Keith & Perry Coal Co., Kansas City, Mo.	45,000
Seattle Automatic Refrigerating Co., Seattle, Wash.	35,000
Amana Society, Amana, Ia.	35,000
Peabody Coal Co., Chicago, Ill.	30,000
Buffalo Smelting Works, Black Rock, N. Y.	25,000
Atcheson, Harden & Co., Passaic, N. J.	20,000
S. G. Parker, Boston, Mass.	20,000
Steinle & Co.'s Brewing Co., Delphos, Ohio	20,000

LOCATION.	DAILY CAPACITY. GALLONS.
Ryan & Richardson, Leavenworth, Kan..	15,000
J. L. Fead, Lexington, Mich.............	15,000
Chappell Chemical Co., Hegewisch, Ill...	12,000
Henry Diston & Sons, Tacony, Pa.......	10,000
Paul Vandenberg Bottling Works, Chicago, Ill.............................	5,000
J. C. Smith Bottling Works, Joliet, Ill...	5,000

BUILDINGS.

LOCATION.	DAILY CAPACITY. GALLONS.
Philadelphia & Reading Railway Terminal Passenger Station.................	500,000
Kansas State Penitentiary, Leavenworth, Kan........................	400,000
Philadelphia Bourse, Philadelphia, Pa....	500,000
Pennsylvania Hospital, Philadelphia, Pa..	250,000
Lorain Apartment House, Philadelphia, Pa.	180,000
Girard Building, Philadelphia, Pa........	180,000
Rialto Building, St. Louis, Mo...........	150,000
Security Building, St. Louis, Mo.........	125,000
Wainright Building, St, Louis, Mo.......	125,000
Murray Hill Turkish Baths, New York..	75,000
Reliance Building, Chicago, Ill..........	60,000
Franklin Sugar Refining Co., Philadelphia, Pa............................	50,000
Chicago Stock Exchange, Chicago, Ill....	50,000
Orthopedic Hospital, Philadelphia, Pa...	30,000
Louise Home, Washington, D. C........	20,000

LOCATION.	DAILY CAPACITY. GALLONS.
W. H. Seymour, Sioux City, Ia..........	15,000
McCahn Sugar Refining Co., Philadelphia, Pa.............	10,000
Spreckles Sugar Refining Co., Philadelphia, Pa.............................	10,000
Erie & Western Transportation Building, Philadelphia, Pa.....................	10,000
J. B. Snook & Son, New York City......	10,000
White Building, Buffalo, N. Y..........	10,000
A. A. Low & Co., New York City.......	10,000
Coburn & Barnum, Cleveland, Ohio.....	5,000
Swift & Co., Chicago, Ill..............	5,000
Fisher Building, Chicago, Ill...........	5,000
Guaranty Building, Buffalo, N. Y.......	5,000

CLUBS, HOTELS, ETC.

LOCATION.	DAILY CAPACITY. GALLONS.
Chicago Athletic Club, Chicago, Ill......	200,000
Young Men's Christian Association, Chicago, Ill..............................	150,000
Manhattan Athletic Club, New York City.	125,000
Midland Hotel, Kansas City, Mo........	125,000
Metropolitan Club, New York City......	125,000
Racquet Club, New York City..........	45,000
University Club, New York City........	30,000
Rittenhouse Club, Philadelphia..........	30,000
Genoa Hotel, Chicago, Ill..............	20,000
Berkeley Club, New York City..........	20,000
Berkeley Lyceum, New York City.......	20,000

LOCATION.	DAILY CAPACITY. CALLONS.
Hotel Walton, Philadelphia, Pa.........	10,000
Wellington Hotel, Chicago, Ill..........	5,000
Logansport Hotel Co., Logansport, Ind...	5,000

ELECTRIC POWER STATIONS.

LOCATION.	DAILY CAPACITY. CALLONS.
Chicago General Electric Transit Co., Maplewood, Ill...........................	600,000
North Chicago St. Ry. Co., Chicago, Ill. .	360,000
Chicago Edison Co., Chicago, Ill.........	300,000
United Electric Light Co., Springfield, Mass......	300,000
Cleveland Electric Illuminating Co., Cleveland, Ohio......................	300,000

RESIDENCES.

LOCATION.	DAILY CAPACITY. CALLONS.
Cornelius Vanderbilt, New York City....	75,000
C. P. Huntington, New York City.......	60,000
A. A. Low, Esq., Brooklyn, N. Y........	20,000
S. V. White, Brooklyn, N. Y.............	20,000
Harry L. Laws, Cincinnati, Ohio........	15,000
C. I. Yerkes, New York City............	10,000
H. R. Hoyt, New York City.............	10,000
John Harlin, New York City............	10,000
George W. Howe, Cleveland, Ohio.......	10,000
William White, Philadelphia, Pa........	10,000
M. Bensinger, Chicago, Ill.............	10,000
F. K. Hipple, Philadelphia, Pa..........	10,000
Benedict Brunswick, Cincinnati, Ohio...	10,000

LOCATION.	DAILY CAPACITY. GALLONS.
Senator Calvin A. Brice, Washington, D. C.................................	10,000
D. F. Keenan, Philadelphia, Pa.........	10,000
Spencer Ervin, Philadelphia, Pa.........	10,000
Mrs. M. M. McCulloch, Philadelphia, Pa.	10,000
Charles Watson, Philadelphia, Pa.......	10,000
C. A. Griscome, Philadelphia, Pa........	10,000
J. M. Townsend, Jr., New York City....	10,000
Alexander M. Guthrie, Pittsburg, Pa.....	10,000
Executive Mansion, Harrisburg, Pa......	10,000

In addition to the above are many other plants, and we herewith append a brief estimate or summary of the various enterprises using our filters and their combined capacity.

SUMMARY
of
JEWELL FILTER PLANTS IN OPERATION.

NAME.	CAPACITY IN GALLONS PER DAY.
Cities	75,000,000
Paper and pulp mills.................	35,000,000
Sugar refineries......................	12,000,000
Mills and factories..................	20,000,000
Public buildings, hotels, clubs, etc.....	5,000,000
Electric power stations...............	3,000,000
Residences...........................	1,000,000
Total...........................	151,000,000

No................   DATE...................

QUESTIONS.

1. What is the source of water supply?

2. What is the size of the supply pipe?   .

3. What is the maximum or fire pressure on pipes?

4. What is the average working pressure on pipes?

Parties desiring informa-   5. What is the filtered water to be used for?
tion or estimate on suit-
able filtering plant will   6. Maximum amount of water required per hour?
confer a favor on us by   7. Average amount of water required per day?
filling out attached blank   8. Is the water generally turbid or roily?
and sending same to us,
together with any other   9. Is the water generally hard or soft?
data or advice which will   10. Is the gravity or pressure system preferred?
acquaint us more fully with   11. Give number and size of filter you prefer to use.
the conditions and require-   12. Have you power to drive agitator during washing?
ments.
                               13. Do you wish price F. O. B. Chicago?

                               14. State maximum capacity of supply pump, if same is used?

                               15. What objectionable features in the water do you wish removed?

                               16. Have you a recent analysis of the water?   .   .

                               17. Have you a storage or receiving tank for filtered water?   .

                               NAME..

                               ADDRESS...........................

JEWELL STILL

STEAM OPERATING

In operating the machines the following general description will apply to the letters shown on the illustration.

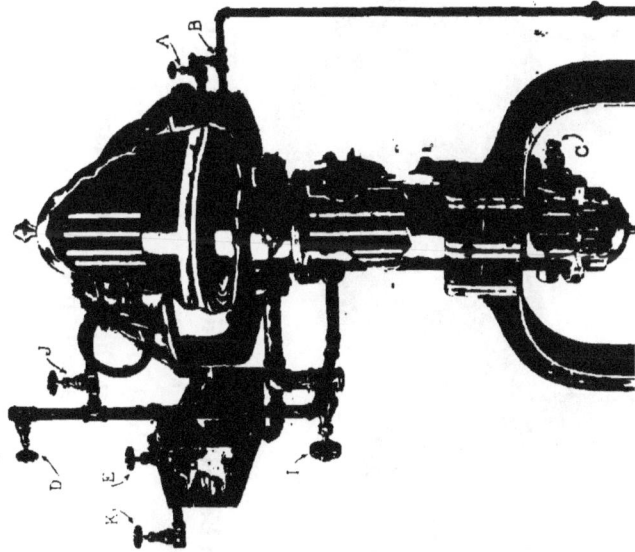

A. Steam inlet to the evaporating coil

B. Steam outlet from the coil to steam trap.

C. Water supply connection.

D. Cooling water discharge.

E. Water supply to automatic feed tank for retort.

F. Distilled water outlet.

I. Blow off from boiling chamber.

J. Hose nozzle for flushing boiling chamber.

connection to boil-
ing chamber.

L. Connection
from evaporating
coil into steam trap.

M. Steam trap
discharge.

N. Steam trap
regulating valve.

Can be installed and operated without a licensed engineer when coil is built for low steam pressure operation.

For this method of operation a low steam pressure boiler is used and connected up so that the coil condensation will return to the boiler, similar to the way a regular return heating system is installed.

Made for high or low steam pressure for the evaporating function of the apparatus. In ordering give steam pressure to be used.

www.ingramcontent.com/pod-product-compliance
Lightning Source LLC
Chambersburg PA
CBHW021938190326
41519CB00009B/1064